SO-AZP-041

STRATEGIC MANUFACTURING

Dynamic New Directions
for the 1990s

PATRICIA E. MOODY
Editor

Dow-Jones Irwin
Homewood, Illinois

© RICHARD D. IRWIN, INC., 1990

Dow Jones-Irwin is a trademark of Dow Jones & Company, Inc.
All rights reserved. No part of this publication may be
reproduced, stored in a retrieval system, or transmitted
in any form or by any means, electronic, mechanical,
photocopying, recording, or otherwise, without the prior
written permission of the publisher.

This publication is designed to provide accurate and
authoritative information in regard to the subject matter
covered. It is sold with the understanding that the
publisher is not engaged in rendering legal, accounting, or
other professional service. If legal advice or other expert
assistance is required, the services of a competent
professional person should be sought.

*From a Declaration of Principles jointly adopted by a Committee
of the American Bar Association and a Committee of Publishers.*

Sponsoring editor: Jim Childs
Project editor: Joan A. Hopkins
Production manager: Carma W. Fazio
Jacket design: Tim Kaage
Compositor: Publication Services, Inc.
Typeface: 11/13 Times Roman
Printer: R.R. Donnelley & Sons, Inc.

Library of Congress Cataloging-in-Publication Data

Strategic manufacturing: dynamic new directions for the 1990s /
  Patricia E. Moody, editor.
     p.   cm.
  Includes index.
  ISBN 1-55623-193-8
  1. United States—Manufactures—Technological innovations.
2. Production management—United States. 3. Manufacturing
processes—United States. 4. Strategic planning—United States.
HD9725.S73 1990
658.4'012—dc20   89-35044
                                     CIP

Printed in the United States of America
1 2 3 4 5 6 7 8 9   0 6 5 4 3 2 1 0 9

for Delia Richard Moody
1917–1972

ACKNOWLEDGMENTS

This project came together with the help and active support of many professional colleagues and friends whose involvement is deeply appreciated. Thanks to:

Steven Wheelwright, who pointed me in the right direction;

Jon Wettstein, (Plant Manager), Elsa Berenberg, Prudence Sullivan, Carolyn Edwards, Roy Leonardi, Ed Bourdeau, Paul Mantos, Bill Mulcahy, Donna McIntyre, Karl Hummel, and Russ Snyder, all of Digital Equipment Corporation;

Kevin Kilbane and Bob Hardison, of Goulds Pumps;

Daniel Holbrook, of the Charles River Museum of Industry in Waltham, Massachusetts;

Erin Murray and Laura Fraser for their very able preparation of one perfect manuscript from twenty different word processing formats;

and finally, Sandy Fleet, Gloria Coolidge, and Doug Glasson, for hustling me along.

Patricia E. Moody

CONTENTS

"Spinning Silk." Boston Manufacturing Company, circa 1890. Reprinted with permission of the Waltham Historical Society.

INTRODUCTION

Twenty years ago I learned a choral piece based on William Blake's "Jerusalem." A few lines from his poem stuck with me all these years, a sad vision of industry gone wrong—England's "dark satanic mills." Here is the original poem.

> And did those feet in ancient time
> Walk upon England's mountains green?
> And was the holy Lamb of God
> On England's pleasant pastures seen?
> And did the countenance divine
> Shine forth upon our clouded hills?
> And was Jerusalem builded here
> Among those dark satanic mills?
>
> Bring me my bow of burning gold!
> Bring me my arrow of desire!
> Bring me my spear! O clouds, unfold!
> Bring me my chariot of fire!
> I will not cease from mental fight,
> Nor shall my sword sleep in my hand,
> Till we have built Jerusalem
> In England's green and pleasant land.

The poem does not simply suggest a yearning to return to the pastoral world we want to believe existed before the mills. I believe it suggests the *potential* for utopia dreamed of by many early innovators (Francis Cabot Lowell and his associates included).

We have come so far since the early days of Lowell's Industrial Revolution. We have managed to leave behind most of the dehumanizing horrors of other industrializing nations—child labor, unsafe conditions, and 12-hour, 6-day work schedules.

In two generations we created a middle class. We have successfully "supplied" two world wars and postwar prosperity.

We in the United States have incredibly rich natural resources and a gloriously diverse heritage of liberated, energized immigrant workers and their descendants.

Our system of government is geared toward generation of wealth. The structure for educating our citizens is in place. And we still subscribe to a humanistic view of man's potential.

In this book the reader will find many approaches to the basic question of how we can compete more successfully. Each section is preceded by an "Overview"; the book is organized so that each chapter stands alone. When read in its entirety, the presentation of manufacturing strategy proceeds from theory and analysis of company and industry through formulation of appropriate competitive strategy and the tactics required to execute that strategy.

We have included a number of very current topics that represent real opportunities for improvement—JIT (just-in-time) for white collar workers, two cases illustrating the empowerment of production workers, time-based competition, and the role of flexibility in manufacturing. The last section addresses the very challenging area of change management.

Patricia E. Moody

PART 1

HISTORICAL PERSPECTIVE

The first chapter of *Strategic Manufacturing* looks at vital statistics describing the comparative strengths of U.S. manufacturing and concludes with an existing approach to filling industry's needs for a better prepared workforce. The section on industry/academic strategic collaboration looks at the need for educational reform to support U.S. strategic requirements in manufacturing and engineering and one university program started by Lester Thurow to answer that need.

Romeyn Everdell traces the evolution of U.S. manufacturing from its beginnings in textiles on the banks of the Charles River ("that Muddy Water") in Waltham, Massachussetts, to Sunnyvale. Two repeating patterns emerge:

- U.S. leadership in innovation began in the early 1800s.
- Each time labor costs became uncompetitive, the industry moved, but the innovators stayed.

CHAPTER 1

VITAL STATISTICS

Patricia E. Moody

Where will the United States be in the 1990s as fierce economic competition becomes the world's focus? Is the United States in its era of final decline or, as more optimistic theorists claim, will it consolidate its strengths—its industry, its tradition of wanting and taking responsibility for global leadership, its diverse and creative cultural energies, its love of money and manufactured "things," its size and natural resources?

The economic future is filled with question marks and surprises. The answers will not come simply internally and, as the United States begins to feel the pressures of being a global player and of having less control over its plans and future, anxiety drives us to want the answers now.

We do have some key pieces of information, vital statistics on leading indicators, that we need to watch and think about over the next three to five years. The data describes and raises questions about the United States' global market share, productivity, competitors' strengths, and U.S. manufacturers' strategic focus. Since *Strategic Manufacturing* went to press, a group of MIT scientists, engineers and economists released the results of their examination of the condition of U.S. industry in their book entitled *Made in America.*[1] From data gathered on eight major industries—computers, commercial aircraft, consumer electronics, steel, chemicals, textiles, autos, and machine tools, they reached conclusions about the problems, including preoccupation with short term returns, lack of cooperation, difficulty translating technology breakthroughs into products, and neglect or misuse of human resources. Among their recommendations are educational reform, and re-structuring government's role in building stronger industries.

SHARE OF WORLD MARKETS

As Table 1-1 shows, the U.S.' share of global markets is challenged by strong competition from Japan, Germany, and Korea. We concede that the United States now holds less of the world auto market; other industry positions are also threatened.

- The U.S. market share of the world car (excluding trucks) market stands at 6.1 percent, Japan at 30.0 percent, Germany at 25.0 percent, and Korea at 1.0 percent and growing.
- In consumer electronics Japan has virtually won the game with total world share of TV and VCR sales of 29.0 percent, and radios/small electronics, 38.0 percent.

TABLE 1-1
World Market Share (1986)[a]

	USA Rank			Japan Rank			Germany Rank			Korea Rank		
	%	Exp	Imp	%	Exp	Imp	%	Exp	Imp	%	Exp	Imp
Auto												
Cars	6.1	6	1	30.6	1	12	24.9	2	3	1.3	11	NA[b]
Trucks	8.8	4	1	36.3	1	NA	13.2	3	6	0.1	22	NA
Chemicals												
Drugs	16.1	2	1	2.5	11	3	16.2	1	2	NA	NA	NA
Plastics	25.4	1	17	4.0	7	6	12.3	3	1	NA	NA	NA
Computers	23.7	1	1	22.0	2	8	10.7	3	2	2.2	11	15
Electronics												
TV, VCR	4.2	6	1	29.4	1	NA	14.9	2	2	NA	NA	NA
Radio, etc.	3.0	10	1	38.3	1	15	4.7	6	2	9.9	2	NA
Household												
Appliances	4.9	7	1	15.9	2	NA	19.7	1	4	5.2	6	NA
Paper	7.5	5	1	3.5	8	8	11.5	4	2	0.5	17	NA
Primary Steel	NA	NA	4	11.4	4	7	19.6	1	2	4.8	8	5
Textiles	5.9	5	1	NA	NA	9	9.1	1	2	4.9	9	NA

[a] World = Non centrally controlled economies.
[b] NA = Country rank is lower than 22 as exporter and/or 20 as importer.

Source: *UN International Trade Statistics Yearbook.*

- The United States' 16.0 percent share in drugs equals the Germans'.
- In computers, the United States holds 23.7 percent against Japan's 22.0 percent and Germany's 10.7 percent. While we may generally assume that American innovative skills will continue to hold the lead, it is alarming to think that the U.S. high–tech industry may go the way of consumer electronics.
- In household appliances, an industry where Black and Decker is making a strong showing worldwide, the United States holds only 4.9 percent, against Germany's 19.7 percent and Japan's 15.9 percent.

Domestic Auto Market

Unfortunately the United States is losing ground in its own backyard. The fight for market share has intensified as each major competitor refuses to cut back production rates; consequently, inventories are building. It has been estimated that there will be five cars for every four buyers by 1990, a good buyer's market, but the overinventoried position sounds a death knell for less competitive car manufacturers and dealers.

- In 1986, approximately 70 percent of U.S. cars were built here by U.S. companies. In 1987 it had slipped to 67 percent. According to a recent MIT study, in 20 years the U.S. has gone from an auto export surplus to an import deficit of $60 billion.[2]
- The individual manufacturers' market shares were divided among three domestic and one Japanese corporation; Chrysler now holds 10 percent of the U.S. market, competing with Ford's 20 percent (growing at a rate of 2 percentage points per year), GM's 36 percent, and Toyota's 10 percent.

FOREIGN COMPANIES IN THE UNITED STATES

As in the auto industry, other foreign competitors have established manufacturing facilities within the United States. The presence of foreign manufacturers, not simply through arm's length ownership from acquisition and buyout deals, is growing. According to the Bureau of Economic Analysis, in 1985 there were over 1500 foreign-owned manufac-

TABLE 1–2
Affiliates of Foreign Companies in U.S. (1985)

	Number	Sales[a]	Inventories[a]
Total	9,824	630,113	64,764
Mfg Total	1,549	185,377	29,322
Auto	25	9,647	1,807
Chemicals	176	62,464	8,916
Computers	37	2,902	NA
Electronics	27	5,028	
Household Appliances	5	1,824	3,483
Paper	37	6,724	780
Steel	51	8,839	3,245
Textiles	78	2,955	621

NA = Not available
[a] in millions of dollars

Source: Bureau of Economic Analysis.

turing facilities operating on U.S. soil, representing sales of $185, 377,000,000. Surprisingly, the largest number of foreign-owned companies manufacture chemicals (34 percent of total manufacturing), followed by autos (Table 1–2).

COMPARATIVE HOURLY WAGES
FOR PRODUCTION WORKERS

We know that one of the causes of high U.S. manufacturing costs (and therefore loss of U.S. competitiveness) is labor rates. The numbers are discouraging.

Clearly, U.S. labor costs have contributed to pricing the nation out of certain markets, specifically the very low (i.e., Hyundai) and very high (i.e., Mercedes and BMW) ends. For example, hourly wages for auto workers in the United States averaged $12.81 in 1986, compared with $8.72 in Germany, $4.00 in Japan, and $2.02 in Korea. Unfortunately U.S. wages, which represent a major portion of total manufacturing costs, have not been changing significantly. In Germany, which holds the second biggest chunk of global auto markets, wages have increased from $6.08 in 1984. The shockingly low figure in Korea can only point to a continued challenge as their auto industry catches up.

TABLE 1–3
R&D Funding as Percentage of GNP

	USA		Japan		Germany	
	Total	Non-Defense	Total	Non-Defense	Total	Non-Defense
1970	2.6	1.6	1.9	1.8	2.1	2.0
1980	2.3	1.8	2.2	2.2	2.4	2.3
1981	2.4	1.8	2.4	2.4	2.4	2.3
1982	2.5	1.9	2.5	2.5	2.5	2.4
1983	2.6	1.9	2.6	2.6	2.5	2.4
1984	2.6	1.8	2.6	2.6	2.5	2.4
1985	2.7	1.9	2.8	2.8	2.7	2.5
1986	2.7	1.8	NA	NA	2.7	2.6

NA = Not available

Source: National Science Foundation.

TRENDS

Research and Development Expenditures

To stay competitive as technological innovators, the United States needs to dedicate more earnings to research and development. Table 1–3 shows that while the percentage of R&D spending as a portion of the U.S. GNP has stayed relatively flat, the two biggest competitors, Japan and Germany, have increased theirs.

As shown in Table 1–4, only 14 percent of total U.S. R&D spending goes to computer development, compared with 12 percent for Japan, and

TABLE 1–4
Percentage of R&D Funding by Industry

	USA	Japan	Germany
Auto	9	15	14
Chemicals	11	20	16
Computers	14	12	14
Electronics (consumer)	22	27	24

Source: National Science Foundation.

14 percent for Germany, even though the United States is being strongly challenged in this area. Also note the heavy Japanese focus on R&D in consumer electronics, an industry they already dominate.

Areas of Competitive Focus

A research report published by Boston University's Manufacturing Roundtable identifies from the responses of 217 industry participants the key areas of competitive focus.[3] In order of importance, these companies rate their need to be competitive as follows:

1. Conformance quality.
2. On-time delivery performance.
3. Quality.
4. Delivery.
5. Speed.
6. Product flexibility.
7. After-sale service.
8. Price.
9. Broad line (features).
10. Broad distribution.
11. Volume flexibility.
12. Promotion.

From this ranking it is clear that quality and cost will continue to be addressed but that the new areas of opportunity are speed, customer service, and flexibility.

Following the rating of priorities, companies developed action lists. In order of importance, quality and statistical process control (SPC) are at the top of the action programs list. It is interesting to note that in 1984 the top three priorities were production control systems, workforce reductions, and supervisor training. Evidently U.S. manufacturers are getting closer to their customers by looking beyond the "mechanics" of production, to issues of quality and responsiveness.

Planned changes in labor costs or in the makeup of the workforce emerge from the survey. Indirect labor will be reduced in the categories of material handling, quality control, supervising, manufacturing management, accounting, finance and administration. Increases are planned for the design and R&D engineering sections, along with direct labor, management information systems (MIS), and purchasing.

Our very comfortable standard of living, supported by high wages, has weakened our competitive position in several industries; the automobile sector illustrates this. Rather than attempting to compete on labor costs, we need to address the issues of improved customer service (from order administration through delivery of quality product), speed, and flexibility. Specific recommendations for improvement in these three key areas are addressed in more detail in subsequent chapters.

INDUSTRY/ACADEMIC COLLABORATION TO IMPROVE MANUFACTURING

The U.S. educational system, from kindergarten through graduate school, is not functioning to produce the numbers and types of skilled workers and managers needed to compete globally. The very fact that foreign languages are not now a general public education requirement is one example. Declining math and science national test scores is another. Although the need for industry/education collaboration is now in the discussion and organization stages, educational reform works slowly and requires strong outside threats to respond. Hopefully President Bush's pledge to be "the education president," and the few innovative programs already begun (including the MIT Leaders in Manufacturing Program) will start the reform.

Engineer Shortage

The Dec. 5, 1988 issue of *Aviation Week and Space Technology* addressed the problem of supply and demand of engineers in emerging technology areas of the U.S. aerospace/defense industry. This industry, which represents a significant amount of the GNP, is being impacted by a shortage of both entry-level, and experienced engineers. It is also one of the two strong U.S. industries (out of eight) according to the MIT Commission on Industrial Productivity. Robotics, artificial intelligence, electronics, systems and software engineering positions are some of the difficult slots to fill.

There are several factors contributing to this problem:

- Demographics—the college population is low.
- East coast companies have difficulty recruiting because of that region's high cost of housing.

- Courses in emerging technologies may not be generally available at the undergraduate level.
- Some experts feel that the current U.S. educational approach limits development in the sense that engineering students may be trained to solve very specific problems, but they are not absorbing what is called "a total systems approach." The need for manufacturing engineering to relate problem solving to all the stages of the product delivery cycle, from engineering through distribution, is just now being accepted and built into curricula.
- The average age of aerospace engineers is increasing. A National Science Foundation Study showed the number of engineers in its over-50 age group increased 23 percent from 1976 to 1986.

Performance Evaluation for the U.S. Educational System

Former U.S. senator Paul Tsongas, who was recently chosen to head Massachusetts' Board of Regents, has proposed a partnership between the public colleges and the high-technology industry. As head of the body governing 29 public colleges and universities, he wants to change current policies to improve the quality of public education. He has proposed the formation of an academic review panel to evaluate candidates for college boards of trustees, the inclusion of business executives on college trustee boards, and the seating of more college professors, presidents, and administrators on corporate boards.

These collaborative ideas grew from his feeling that the dominant need is to ensure the survival of the American standard of living and quality of life by achieving true competitiveness in international trade. To meet this challenge the United States needs a superbly trained, educated, and motivated workforce. The United States is competing with the Pacific Rim nations and Europe 1992, both of which trade entities have stronger, more effective educational systems.

One of Senator Tsongas' specific recommendations to improve the quality of the U.S. educational processes is the introduction of performance-based funding allocations—in his words, "Excel and you'll get more funding; botch it and we'll cut your funds."[4] The issues of accountability and quantitative performance evaluation are quite clear in the private sector; he wants to see them "revisited" in the academic world. Politics tends to interfere with public higher education. The Senator believes the powers-that-be should agree to a merit-based philosophy of decision making within the institutions, as well as agree on the larger issues of the missions of individual institutions and their structure. This means that both the business and academic communities need to agree on the educational requirements that will best satisfy our need to compete in international business.

Educational reform is driven from enlightened self-interest, in which each sector sees it own success and/or survival linked to the success and/or survival of each other sector. The following experimental education program is very specifically geared toward satisfying these needs.

Senator Edward Kennedy, chairman of the Senate Labor and Human Resources Committee, says that, "Unless we act now, America will soon be without enough hands on deck to get the work of America done."[5] A Labor Department study has predicted a problem caused by a surge in new jobs and a decline in the birth rate, compounded by what the Senator calls a potential "skills gap" between available workers and job openings. Collaboration between industry and education can provide the initial energy to start reforming U.S. education and training to overcome these problems.

The MIT Leaders in Manufacturing Program— Developing a New Type of Manager

The Leaders for Manufacturing Program is a new, experimental, five-year, joint educational and research program established to help the United States recapture world leadership in manufacturing. It represents an exciting collaboration between the MIT Schools of Engineering and Management and eleven of the nation's leading manufacturing corporations. The program is funded by a $30 million collaboration between the university and eleven companies—Alcoa, Boeing, Digital, Eastman Kodak, Hewlett-Packard, Johnson & Johnson, Motorola, Polaroid, General Motors, Chrysler, and United Technologies.

The goals of the program are to develop a model curriculum for educating a new generation of leaders in manufacturing, and to draw some of the nation's best students into careers in manufacturing industries. (See recommendations of Romeyn Everdell in Chapter 2).

Lester Thurow, the Dean of the Sloan School of Management at Massachusetts Institute of Technology, feels that their new combination management/engineering graduate degree program is exciting and necessary. A common criticism of MBA programs is that they do not adequately address the type of training managers will need to lead U.S. industry. According to Thurow, "Today, most top managers in Europe and Japan have a technical education, while most U.S. managers don't. As a result, foreign companies have been more willing to take risks on new technologies."[6]

Professor David Hardt of the MIT School of Engineering agrees with Cohen and Zysman (authors of *Manufacturing Matters, the Myth of the Postindustrial Economy*) that it is manufacturing, not the service sector only, that creates wealth. It is misleading and dangerous to think that the United States can safely abandon basic production and become fast-food purveyors to the world.

Curriculum

The two-year program includes academic course work, and six months of on-site research experience with a sponsoring organization under close faculty and industrial supervision. The program's purpose is to define the combination of educational experiences that will yield graduates who can be measurably more effective in the definition, design, manufacturing, and delivery of high-quality products and systems.

The Fellows, as the degree candidates are called, all take a management core curriculum as well as selected electives. Within the School of Engineering, they design individual programs in an engineering department. In addition, all students take a required weekly Professional Seminar, the Program's major mechanism for bringing together its disparate constituencies of management and engineering faculty and students and company sponsors. During the seminar, students and faculty also look closely at each of the sponsoring companies, with student teams developing presentations and companies bringing in their best in-house instructors.

The program represents a commitment to integrating engineering and management concerns. Engineers and managers are encouraged to learn more about each other, to begin to speak the same language.

An example of the very intensive, but typical, two-year curriculum for a dual degree includes thirteen management and six engineering courses, four seminars, three projects, and four theses.

Graduates are expected to leave the program with many of the following:

- A broad understanding of manufacturing, from the product design concept to its production, use, and field maintenance.
- An understanding of production planning, scheduling, inventory control, manufacturing policy, and management accounting.
- An understanding of the role that human factors (e.g., labor relations, psychology, and ethics), logistics, marketing, vendor support, and economics play in all aspects of manufacturing systems.
- Self-confidence in approaching the design and management of innovative manufacturing and processing systems, and a tolerance for ambiguity during times of change.
- An in-depth understanding of at least one technology and its integration into an overall system (e.g., mechanical assembly or integrated factory information systems).
- An understanding of materials processing technologies and methods of modeling, simulating, and controlling processes.
- Working understanding of computer-based design, manufacturing system automation and control, and communications systems.
- An understanding of emerging technologies and materials that will impact manufacturing beyond the next decade.
- Increased leadership strengths and improved management, communication, and interpersonal skills.

The Students

The first 20 fellows were 12 of the top applicants to the Engineering Graduate School and 8 of the top applicants to the School of Management. The current students were offered grants of full tuition (the cost of one year at MIT is $25,000) and a monthly stipend of $900 for two years.

This program represents a new educational model in the area of university/manufacturing alliances, as well as interdepartmental collaboration. Professor Bitran of Management Science Department declares, "We want to be copied. And we want companies to count on the program's graduates."[7]

GLOBAL COMPETITION

Four years ago the President's Commission on Industrial Competitiveness completed a two-volume study with specific recommendations on how to meet U.S. global competitive challenges. The commission included leaders from industry, the academia, banking, government, and various professional organizations. In *Volume I, The New Reality*,[8] the recommendations included:

In Research and Development and Manufacturing
- Create a Federal Department of Science and Technology.
- Increase tax incentives for research and development.
- Commercialize new technologies through improved manufacturing.

In Human Resources
- Increase effective dialogue among government, industry and labor.
- Build labor/management cooperation.
- Strengthen employee incentives.
- Deal with displaced workers.
- Improve work force skills.
- Improve engineering education.
- Improve business school education.
- Create partnerships in education.
- Develop education technology.

In International Trade
- Minimize impact of controls on competitiveness.
- Export expansion.
- Export financing.

How many recommendations have been acted upon four years later? Although we are working on strengthening our trade position, most activity is happening in the Human Resources category; the first group of recommendations on R&D and manufacturing remain relatively untouched.

Europe 1992

A big area of confusion for U.S. manufacturing is the 1992 creation by the European Community of a new super trading partner. U.S. companies already manufacturing in Europe, as well as those exporting to the member nations, are looking for answers. There is great uncertainty over what to expect and how to compete. Indeed, the Commerce Department has set up a new office to assist the transition, called the Single Internal Market Information Center, in Washington, D.C.

Adapting to the impending European trading accord is not unlike the process of changing U.S. tax laws. The first step, passing legislation to revise the tax code, did not immediately affect the taxpayer. The tax laws, the working regulations that we need to complete our returns, are actually written as we use and test the guidelines. The gaps are filled in as precedents are set, and new instruction books compile more helpful information.

The 1992 accords are actually an array of directives, over three hundred specifics, designed to remove physical, fiscal, the technical barriers to trade among twelve nations (Belgium, Denmark, France, West Germany, Greece, Ireland, Italy, Luxembourg, Netherlands, Portugal, Spain, U.K.) making up the European Community (EC). Although we cannot specify all the changes the new regulations may require of U.S. manufacturers in Europe, we know that the Europe 1992 agreements effectively create the world's third major trading block (illustrated in Table 1–5).

Joint Research and Development

According to a report of The Conference Board entitled *1992: Leading Issues for European Companies*,[9] there is an EC-administered $180 million fund dedicated to promote cooperation among research labs of member states. This combination of R&D resources may, particularly

TABLE 1–5
Three Global Trading Blocks

	U.S.	Japan	EC
Population	243.8M	122.0M	323.6
Share of world GNP	25.9%	9.4%	22.1%
Total labor force	121.6M	60.3	143.0
Agricultural	3.4	4.6	11.9
Nonagricultural	118.2	55.7	131.1
GNP Growth Rate 1987	2.9%	4.2%	2.9%
Exports (US$ millions)	250.4	231.2	953.5[a]
Imports (US$ millions)	424.1	150.8	955.1[a]

[a] Includes trade between EC members.

Source: *The Wall Street Journal*, 2/13/89 and 1/23/89.

in the electronics field, enable the Europeans to beat Americans and Japanese to market. The first phase of the joint research and technology programs, Esprit, was launched in 1984, scheduled to run 10 years. Part One of Esprit involved 450 participating organizations. Part Two will deal with advanced microelectronics, software development, advanced information processing, office automation, and robotics. Later phases will explore new industry technologies such as CAD/CAM.

Regulations Impacting U.S. Manufacturers
The accords will affect how U.S. manufacturers do business in Europe. Technical standards vary from one European country to another; French and Italian electric plugs, for example, have three prongs, while West German plugs have two. Hopefully, the rules will be standardized, along with manufacturing safety standards and other product specifications. Other areas included for standardization ("harmonization") are:

- Motor vehicle emission standards.
- Measuring instruments.
- Machine safety.
- Food additives.
- Noise levels in lawn mowers, household appliances, and hydraulic equipment.
- Pharmaceuticals, high-tech medicines and certain devices.
- Labor issues (which may increase worker mobility within Europe): engineering training standards, comparability of vocational training qualifications, and mutual recognition of higher educational diplomas.

Transport time, and therefore customer delivery times, should be reduced as border checks of paperwork are simplified.

SUMMARY

We have briefly reviewed the factors impacting U.S. manufacturers' ability to compete globally, including:

- The loss or threatened loss of market share.
- The "invasion" by foreign manufacturers.
- Research and development expenditures.

• Internal focus on quality, flexibility, and customer service.
• The need for education/industry collaboration.
• The 1992 EC reforms.

Of all the issues faced, U.S. manufacturers can really only address four directly: labor costs, R&D spending, internal manufacturing strategic focus, and educational reform. But effective focus on even one will realize some of the potential for productivity gains.

ACKNOWLEDGMENTS

Thanks to Paula Cronin, of *MIT MANAGEMENT,* who contributed her piece on the Leaders in Manufacturing Program, to Dean Lester Thurow and Professor Don Rosenfield of the MIT Sloan School, also to Paul Tsongas for their very generous and timely contributions to this chapter.

Research describing market share and other key data was performed by Monique Richardson of the Simmons College Graduate School of Management Library.

NOTES

1. Michael L. Dertouzos, Richard K. Lester, Robert M. Solow, and the MIT Commission on Industrial Productivity, *Made in America.* Cambridge, MA: The MIT Press, 1989.
2. Ibid., p.18
3. Jeffrey G. Miller and Aleda Roth, *Manufacturing Strategies, Executive Summary of the 1988 North American Manufacturing Futures Survey,* Boston University School of Management Manufacturing Roundtable, 1988.
4. Quoted from Senator Tsongas' speech before the Massachusetts High Tech Council, February 1989.
5. Boston *Globe,* January 30, 1989, p. 31
6. Boston *Globe,* June 19, 1988, p. 97.
7. Alfred P. Sloan School of Management, *MIT MANAGEMENT.* Cambridge, MA: Massachusetts Institute of Technology, Winter 1988, pp. 2–8
8. United States Presidential Commission on Industrial Competitiveness, *Global Competition: The New Reality.* Washington, DC: The Governement Printing Office, 1985, pp. 45–46.
9. Catherine Morrison, *1992: Leading Issues for European Companies.* NYC: The Conference Board, 1989.

STATISTICAL REFERENCES

Arpan, Jeffrey S. and Ricks, David A. *Directory of Foreign Manufacturers in the United States*, 3rd ed. Atlanta: Georgia State U., 1985.

Duns Analytical Services. *Industry Norms and Key Business Ratios*. Murray Hill, NJ: Dun & Bradstreet, Inc., 1983–84 through 1987–88 (annual).

Industry Surveys. New York: Standard & Poor's, 1988.

International Labor Office. *Yearbook of Labor Statistics*. Geneva: International Labor Office, 1987.

International Monetary Fund. *International Financial Statistics Yearbook*. Washington, DC: 1987.

National Science Foundation. *International Science and Technology Update*. Springfield, VA: NTIS, 1987.

————. *Research and Development in Industry, 1984*. Springfield, VA: NTIS, 1986.

"100 Leading National Advertisers." *Advertising Age*, Sept. 28, 1988; (59) 41.

Pennar, Karen. "The Factory Rebound May Be More Fantasy Than Fact." *Business Week*, 3083:98, Dec. 12, 1988; 101

"Projections 2000." *Monthly Labor Review*. 110 no. 9 (Sept. 1987), pp. 3–62.

RMA Annual Statement Studies, FY 1988. Philadelphia: Robert Morris Associates, 1988.

Rosenbaum, Andrew. "Europe: Fortress—or Facade?" *Industry Week*, 238 no. 3 (Feb. 6, 1989) pp. 54–55.

"Special Report: Reshaping Europe: 1992 and Beyond," *Business Week* 3083 (Dec. 12, 1988), pp. 48–73.

Survey of Current Business 68 no. 7. (July, 1987).

U.N. Department of International Economic and Social Affairs. Statistical Office. *Industrial Statistics Yearbook 1985*. New York: 1987.

————. *International Trade Statistics Yearbook 1986*. New York: 1988.

U.S. Bureau of Labor Statistics. *Handbook of Labor Statistics 1983*. Washington, DC: GPO, 1985.

U.S. Bureau of the Census. *Annual Survey of Manufacturers*. Washington, DC: GPO, 1983 through 1986 (annual).

————. *Statistical Abstracts of the United States*. Washington, DC: GPO, 1988.

U.S. Dept of Commerce, Bureau of Economic Analysis. *Direct Investment Abroad: Operations of U.S. Parent Companies and Their Foreign Affiliates, Preliminary 1986 Estimates*. Washington, DC: GPO, 1988.

————. *Foreign Direct Investment in the United States: Operations of U.S. Affiliates, Preliminary 1985 Estimates*. Washington, DC: GPO, 1987.

U.S. Dept of Commerce, International Trade Administration. *U.S. Industrial Outlook*. Washington, DC: GPO, 1988.

U.S. National Center for Education Statistics. *Digest of Education Statistics*. Washington, DC: GPO (Dept. of Education. Office of Education Research and Improvement CS88-600), 1988.

————. *Projections of Education Statistics to 1997–98*. Washington, DC: GPO (Dept. of Education. Office of Education Research and Improvement CS88-607), 1988.

UNESCO *Statistical Yearbook*. Paris: UNESCO, 1987.

GPO: Government Printing Office
NTIS: National Technical Information Service

CHAPTER 2

FROM LOWELL TO SUNNYVALE: MANUFACTURING IN THE UNITED STATES

Romeyn Everdell

Editor's note

As the United States approaches the last decade of the 20th century, large segments of U.S. industry ("smokestack America") are facing crises. The question remains: can U.S. manufacturers compete in basic manufacturing on a world-wide basis, or is U.S. industry over the hill? Is the U.S. competitive position like that of a professional athlete weakened by age, unable, like the United Kingdom, to stay in the game? Or can industry be revitalized in a competitive response to the challenges from Japan, West Germany, and now South Korea? This book takes the position that industry is a team sport. With fresh players, updated strategy and training, and new management, it is not only possible to regain the competitive edge; we actually know how to do it.

There is an old but compelling cliche that those who fail to study history are condemned to repeat the mistakes of the past. We have neither the time nor resources to repeat mistakes. It is important, therefore, to review the evolution of manufacturing before addressing the current challenge and response.

MANUFACTURING BEGINNINGS

Manufacturing had a noble beginning, and in spite of traumatic events in its uneven progress, we recognize it was man's tool-making ability

that distinguished humans from their animal ancestors and built civiliza-
tion as we know it. From man the *faber* (Latin for "maker") gradually
evolved the artisans and craftsmen of the middle ages. The Medieval guilds and shops were our first manufacturing facilities. However, *modern* manufacturing received its initial impetus from the
Renaissance. The Age of Reason focused on the mind rather than the
spirit and fueled the explosion of science and engineering. Leonardo
da Vinci was the first in a long line of technical innovators: Edison,
Bell, Ford, and the current giants such as Shockley, Packard, and
Jobs.

Modern manufacturing as we know it began in England between
1765 and 1815 with the Industrial Revolution—a child of the marriage
of emerging technology and the handcrafts. Manufacturing became a
major influence for good—availability of low-cost products—and evil:
"the dark satanic mills" that led to social stress, labor/management conflict, and the still-to-be-resolved communist/capitalist schism. Modern
manufacturing, with its origins in human labor (the worker) and human
mental skills (the engineer), has been and still is a major factor in defining the quality of life of society and is therefore worthy of our interest,
understanding, and involvement.

U.S. MANUFACTURING MILESTONES

Wickham Skinner suggests in a chapter he wrote entitled "The Taming
of Lions: How Manufacturing Leadership Evolved 1780-1984", from
the book *The Uneasy Alliance*, that there are five stages to the evolution
of manufacturing in the United States:

1800–1850	The Age of the Technical Capitalists
1850–1890	The Introduction of Mass Production
1890–1920	Scientific Management in Manufacturing
1920–1960	The Golden Years of U.S. Manufacturing
1960–1980	The Decline of the American Factory

Basically these milestones identify responses to evolving technologies, changing markets, proliferation of products in an increasingly
wealthy nation, and intense competition (see Figure 2–1).

FIGURE 2–1
Evolution of Manufacturing in the United States

TIME PERIOD		1800	1850	1890	1920	1960	1980	
STAGES	RENAISSANCE TO INDUSTRIAL REVOLUTION	TECHNICAL CAPITALISTS	MASS PRODUCTION	SCIENTIFIC MANAGEMENT	GOLDEN YEARS	DECLINE	CHALLENGE FROM OVERSEAS	
KEY PLAYERS	LEONARDO DaVINCI TO CARTWRIGHT	FRANCIS CABOT LOWELL WATT	BELL EDISON	TAYLOR FORD	KNUDSEN CHARLIE WILSON SHOCKLEY			
FACTORIES	• CRAFTS • GUILDS • COTTAGE INDUSTRIES	• WATER POWER • LARGE LABOR FORCE • TECHNICAL INNOVATION	• CONVERSION TO STEAM AND ELECTRICAL POWER • HIGH VOLUME • INCREASING AUTOMATION	• PRODUCT PROLIFERATION • JOB SHOPS APPEAR • REPLACEMENT OF LABOR WITH MACHINERY	• RAPID PRODUCT OBSOLESCENCE • FURTHER REDUCTION OF LABOR THROUGH AUTOMATION • HEAVY INVESTMENT IN EQUIPMENT	• INFLEXIBLE FACTORIES • EQUIPMENT OBSOLESCENCE		
MANUFACTURING MANAGEMENT	CRAFTSMEN TO APPRENTICE	• FACTORY AGENT	• AGENT PLUS FOREMAN	• INCREASING PRODUCTION STAFF • PRODUCTION MANAGER	• COMPLEX PRODUCTION STAFF. HIGH OVERHEAD	• ATTEMPT TO INCREASE CONTROL AND USE COMPUTERS		

• PRESIDENT INTIMATELY INVOLVED • SMALL STAFF OF SUPERVISORS • AGENT RUNS MILL	• PLANT MGR W/ FOREMEN • TOP MANAGEMENT IN TECHNICAL INNOVATION & EQUIPMENT INVESTMENT	• FOREMEN LOSE POWER TO PRODUCTION MGR • TOP MANAGEMENT LESS INVOLVED	• LARGE SPECIALIZED STAFF • FACTORY RUN BY NUMBERS	• BUREAUCRACY • MFG ISOLATED, RUN BY NUMBERS
• COMPETITION BECAUSE OF SUCCESS	• COMPETITION FROM LOW WAGE REGIONS	• COMPETITION INCREASINGLY RESULT OF ECONOMY OF SCALE	• COMPETITION BASED ON CUSTOMER SATISFACTION • LARGELY WITHIN U.S.	• COST COMPETITION FROM OVERSEAS

ORGANIZATION

COMPETITION

Summary of U.S. Manufacturing Trends

As the nation grew and prospered, more U.S. assets were used to develop a manufacturing base. While the jobs created by this investment and continuing wage increases improved the living standard, over time there was also a gradual but dramatic reduction in direct labor per unit of product. The reduction in direct labor content was achieved by a growing production staff of industrial engineers, planners, controllers, until a bureaucracy of manufacturing middle management developed that increased overhead costs and partially offset the reduction in direct labor. Paralleling this trend came the need for top management to give greater attention to selling and marketing, accounting and financing, and legal skills to meet competition in an increasingly regulated environment and to keep the investment in production fully utilized.

Manufacturing in the 1800s was the only game in town. By 1980, largely due to early successes, it had fallen to low man on the totem pole on the stack of management concerns. To put flesh on these broad trends and to better understand the implications and impact of the current dilemma, let us look at each of Skinner's five periods in more detail.

1800–1850, THE AGE OF THE TECHNICAL CAPITALISTS

The Boston Manufacturing Company

Prior to 1800, the United States was a nation largely of farmers and maritime merchants. Much of the demand for manufactured goods was met by imports shipped from Europe, India, and the Orient. As a result, significant amounts of capital started to accumulate in the hands of New England's merchants and shipowners. One important import was machine-woven cotton textiles from England, a major product of the British Industrial Revolution. In 1811 a successful Boston merchant, Francis Cabot Lowell, went to England, visited Manchester, the center of that revolution, and in an astonishing act of memory became the first U.S. industrial spy. He returned to Boston with mental blueprints of the power loom, a proprietary British invention. On his return, using his own and additional capital from his association with fellow maritime merchants, Lowell formed The Boston Manufacturing Company and built the first American powerloom textile mill in Waltham, Massachusetts, in 1813. With the help of a brilliant mechanic, Paul Moody, Lowell adapted and expanded the technology taken from England to produce a completely integrated manufacturing process at a single site, starting with raw cotton and ending with finished cloth.

In one extraordinary leap, Lowell created a stock corporation, formed an integrated manufacturing facility with a significant competitive edge, and hired the first manufacturing engineer. Lowell's enterprise brought the stockholders between 1817 and 1821 just over 19 percent average annual return on investment. In 1822, as the company grew, the return was a tax-free 27.5 percent. The original Boston Manufacturing Company of Waltham expanded to the banks of the Merrimack River in Lowell, Massachusetts. Competitors were quick to develop, and many more mills followed throughout New England wherever water power, the required energy source, was available. A fascinating history of these beginnings and their impact on American society can be found in *The Enterprising Elite—The Boston Associates and the World They Made.*[*]

Key Developments of the Period

It is important to recognize that the initial development was technology. The acquisition of capital to realize the potential was vital but it was the second step. The implication of the word *capitalism* tends to obscure this fundamental principle.

In Lowell's original organization, three items stand out: (1) management structure, (2) sales, and (3) the labor force. The early mills were run by a mill agent who managed the day-to-day business. The agent reported to the treasurer, who either was or reported to the president. The treasurer did purchasing as well as handling all financial affairs. Sales were achieved by a separate commercial firm paid commissions. This activity also fell under the control of the treasurer. As a result, there was very little overhead to spread over a large manufacturing labor base.

Despite all the technological innovations in carding, combing, spinning, and weaving, a textile mill of this period had a huge labor force. In the mid-1800s labor represented in excess of 40 percent of cost-of-sales (at Amoskeag, the largest textile mill in the world, wages were still 35 percent of gross income by 1900). The labor force was of great concern to Lowell due to his reaction to the dreary factories of the British Industrial Revolution. Charles Dickens' novels vividly portray the social

---

[*]For this and other references, see Bibliography at end of chapter.

consequences of the Industrial Revolution in England: Factory profits made the rich richer, while subsistence wages created appalling living conditions for labor in the factory cities. Lowell copied only the English loom design and evolved an ingenious solution to the labor problem. He devised the concept of hiring young unmarried women off the farm, training them, and acting *in loco parentis*. They lived in mill-owned, supervised housing. As a social engineer, Lowell viewed his effort as improving the limited opportunities available to women off the farms. The "singing loom girls," whose ethics and behavior became the concern of the benevolent mill owner, were his answer to the wretches laboring in the Manchester mills. This interesting experiment did not last, but at least the beginnings of the U.S. factory system were both innovative and progressive.

Between 1820 and 1850, the success of The Boston Manufacturing Company stimulated growth in other textile mills, and Lowell started the capital goods industry through sale of textile machinery. Arms manufacturing, another seminal manufacturing activity, developed the concept of interchangeable parts and opened the door to mass production. Canals, followed by railroads around 1830, reduced the cost of transporting raw materials and finished products to and from manufacturing facilities and expanded the impact as well as the market for these factories.

At the end of the 18th to the middle of the 19th centuries, manufacturing began to rival farming and maritime shipping. The new game in town attracted a nucleus of wealthy, technically oriented entrepreneurs. They were interested in and involved with manufacturing equipment and process and wanted to deal intelligently with the new industrial work force. The early managers gave manufacturing their full attention.

1850–1890, INTRODUCTION OF MASS PRODUCTION

Evolution of the Factory and the Work Force

The Civil War and migration of farming to the West changed the isolated Eastern factory towns to an urban industrial region. Technological restraints were removed by the invention of energy and communications devices—the steam engine, telegraph, electricity, and the electric motor. Manufacturing broke free of location restraints; plants could be located close to raw material, markets, or labor. Factory wages attracted the subsistence farmers and stimulated demand for manufactured products. Technology was applied to both new products and the manufacturing process. The era of mass production arrived.

Labor became available as agriculture receded and emigration from depressed regions in Europe increased. The early stage of automation reduced costs dramatically by trading investment in manufacturing equipment for reduction in direct labor. Production output increased. Although new products came into being, variety within a product was limited. Each plant produced relatively few products. Top management focused on economies of scale. Managers were directly involved with production technologies and concerned with gaining a satisfactory return on the escalating investment in manufacturing equipment.

Lowell's "noble experiment," founded on an enlightened labor policy, deteriorated. It was too costly to train the transient flow of inexperienced farm girls. A permanent work force was the cost-effective answer. As high profits created competition and pressure on prices, the need to reduce labor costs, coupled with the freedom of plant location, started the search for low-cost labor regions—initially the post-Civil War South. As the permanent work force settled in, the social mores of the male breadwinner and the physical demands of many of the new jobs created an increasingly male-dominated factory.

The financial requirement to keep expensive production machinery fully utilized under growing competition put pressure on labor, the major variable cost, particularly in periods of depression. Top management delegated all responsibility for the labor force to the agent and he in turn to the foreman. The foreman was king, responsible for hiring, firing, directing, and controlling labor. The reaction of the foremen to pressure for increased output inevitably led to labor problems and helped develop and expand unionism. However, business prospered, labor requirements grew, and industrial urbanization continued.

The Amoskeag Manufacturing Company— A Case in Point

The textile industry was a leading indicator in the manufacturing community. Hareven and Langenbach's *Amoskeag, Life and Work in an American Factory City*, details the fascinating history of a textile factory located in Manchester, New Hampshire, across three periods of manufacturing history. Amoskeag was a direct descendent of the

Waltham-Lowell system. In the mass production era it became the largest cotton textile mill of its time. It declined as textiles moved south.

Although the mill was started along the banks of the Merrimack River, the automatic loom gradually replaced Lowell's water-powered loom. The process was technologically driven and became very successful because of the application of mass production techniques. The labor force, which reached a high of 17,000 workers, consisted of New England mill girls, later replaced by French Canadians and Europeans. The earliest agent was a model of corporate paternalism, but the benevolence of his regime deteriorated as size, makeup, and competitive pressures impacted the labor/management relationship. Finally, the seeds of decline grew as new mills opened in the South with cheaper labor and equipment better suited to the changing product. But until the end of the 19th century Amoskeag was a very profitable model of manufacturing of that era.

General Characteristics of the Period

The period of mass production was a logical extension of the 1800–1850 origins. Sales were still largely conducted through independent companies: management was financially and technologically oriented; and top management, a small group generally located away from the plant, consisted of a president (who sometimes doubled as treasurer) and a slowly growing staff of clerks. Manufacturing overhead grew as size and complexity required greater numbers and layers of foremen, superintendents, and department heads, with the agent at the top of this expanding pyramid. Technology also required expansion of the indirect labor force involved with maintenance, material handling, inspection, and payroll-related duties. During this period of large direct labor forces, however, indirect labor costs were insignificant.

**1890–1920, SCIENTIFIC MANAGEMENT
IN MANUFACTURING**

Enter the Consultants

At the turn of the century, outsiders (the first consultants) began to impact all industry. The innovators included Frederic Taylor, Frank Gilbreth,

Henry Gantt, Charles Bedeaux, and Lyndall Urwick, who fathered a more systematic approach to manufacturing. Their new techniques came to be known as "scientific management," a somewhat misleading term for the origin of industrial engineering.

The age of mass production had generated a group of operational concerns:

- Personnel problems—hiring, training, and integrating labor into an increasingly complex manufacturing process.
- Manufacturing or process engineering—the design of production equipment to suit product and volume.
- Rate setting—the determination of the expected number of pieces to be produced per worker and piecework pay rates.
- Material control—the management of logistics from raw material through work-in-process, storage, and finally shipment.

The Birth of Industrial Engineering

The initial efforts of people like Taylor and Gilbreth focused primarily on reduction of direct labor costs through methods and time study. The systematic approach of these pioneers required the addition of a staff of industrial engineers. The valid assumption was that a dollar spent on trained staff would realize two dollars of direct labor savings. They formally analyzed methods used to run production equipment or perform manual tasks, improved that method, compared before and after with a stop watch to determine the degree of improvement, and computed the savings. The contribution of Bedeaux was to take this time data and establish a "standard hour" for each direct labor operation based on the number of pieces that could normally be produced in 60 minutes. He used this time standard as a basis for standard hour incentive pay plans to replace the piecework plans set rather arbitrarily by foremen.

The industrial engineers reduced labor; increased output; and, by distributing the savings to the worker as well as the company, increased workers' wages. A byproduct of these productivity programs was to dramatically reduce the foreman's power by taking away his ability to determine workers' pay. Even setting the base rate (paid for standard performance or below) became the responsibility of the personnel department, who were influenced by area wages.

Growth of Production Staff

Another important byproduct of industrial engineering was the use of the standard hour as the basis for developing the standard cost of labor and estimating product costs and prices. The labor standard also became a control tool, used to compare actual to standard and to generate variances. The opportunity to monitor labor performance led to the addition of whole departments of cost controllers. The manufacturing arm soon grew trained staffs of salaried specialists gathered into production departments:

- Methods and time study.
- Cost accounting.
- Personnel.
- Material control.
- Manufacturing engineering.

Most of these functions had originally been part of the foremen's responsibilities. Cost accounting expanded the skill level and workload of the corporate clerk, a one- or two-man crew in the 1800s. Manufacturing engineering evolved from the relatively few master mechanics and inventors who had designed the automated equipment on which the Industrial Revolution was built. The number of manufacturing engineers grew as new technologies became available to accelerate the replacement of manual labor with machines. Industrial engineers ("efficiency experts," often accused of speed-up) tried to apply labor more efficiently by reducing the time required by the remaining workers to produce product. Management began to use the new standards to look for unfavorable variance from direct labor standards. A significant change emerged as top management moved away from direct involvement with manufacturing toward a delegation of responsibility to a growing production management staff, while monitoring quantitative performance measures (management by objectives or MBO).

As important as was the impact on all industry of the "systems engineers," the activities of specific manufacturers were equally important. Under the stimulation of technological giants like Edison and Bell, new products and new industries vastly expanded the manufacturing base. Among these facilities were a few model factories (e.g., the Hawthorne Works of Western Electric in Chicago) that applied all the new "scientific management" techniques to manning high-speed machinery. This period saw the introduction of Henry Ford's moving assembly line and the culmination of his manufacturing masterpiece: the River Rouge complex in Dearborn.

Managing Labor

There was a growing concern that mass production had been achieved at the expense of the worker. The emerging dominance of an engineering view of the world, fueled by the popular belief that progress equaled improved technology, was changing the production labor force from man the *faber* to man the machine. The most biting commentary on this change was the Charlie Chaplin film *Modern Times*.

Unions had been hard at work to ensure the worker a financial share in the industrial revolution, but unrest in the labor force went deeper. Experiments at the Hawthorne Works were the first efforts to apply industrial psychology to the monotonous and dehumanizing effect of tasks broken down into small, repetitive elements. The Hawthorne experiments seemed to suggest that monotonous, unskilled, repetitive jobs could be accommodated if the foremen showed more concern for the human or spiritual side of the worker, suggesting that the Age of Reason's rationality had gone too far.

Benevolent experiments built on paternalism were tried at Draper, Hershey, Kohler, and others. They all foundered on the "owners-know-best" restraint on the freedom of the individual, complicated by the diversity of an increasingly pluralistic and educated work force. Labor unrest continued while well-intentioned efforts to deal with it generally deteriorated into an adversarial relationship between union and management.

Impact of Customer Demands

Important manufacturing strategy changes in this period were influenced by the proliferation of not only new products but product variety. The Ford manufacturing ideal of "any color as long as it is black" did not last. Marketing became a vital concern driving product design and manufacturing. The use of separate sales companies proved uneconomical and too remote. As a result, top management gave more attention to the creation and management of specialized internal departments for sales and marketing. Concern with marketing accelerated the drift of top management away from exclusive interest in manufacturing and finance. At the same time, manufacturing faced increasing problems dealing with product change and proliferation in plants originally designed for simpler and often dramatically different items.

The Decline of Amoskeag—A Harbinger of the Future

Through the advantage of hindsight, the history of Amoskeag in this period is a remarkable forerunner of what occurred 50 years later throughout U.S. industry. The principal difference is that competition came from the low-cost labor American southland rather than off-shore. New and more advanced textile factories with lower labor rates created cost competition. Amoskeag reacted by going from one-shift to a three-shift, seven-day schedule. Tripling production capacity did not help the saturated textile market. Prices fell and profits dropped.

With lower margins, management was reluctant to invest to upgrade the manufacturing process. They chose instead to run fully depreciated and rapidly deteriorating machinery at higher rates for longer hours. Management reduced labor rates with arbitrary piece rate changes. Labor/management relations worsened. By 1920 the surge of demand to supply the U.S. troops in World War I with uniforms, blankets, and the like, masked the symptoms of decay. As a result, the problems that would bring about the tragic ending of this once noble enterprise by the mid-1930s were not addressed.

1920–1960, THE GOLDEN YEARS OF U.S. MANUFACTURING

World War I Aftermath

U.S. Manufacturing entered a period of world domination at the end of World War I. The Versailles Treaty levied heavy reparation payments on the defeated nations, which had a depressing effect on world markets. When the pent-up domestic demand of the war years was satisfied, the U.S. economy went into a tailspin in 1929, triggering a prolonged world-wide depression throughout the 1930s. This period forged the crucible that carried the United States into the golden era of manufacturing after World War II. In the 1930s, management had to cut back to survive.

With labor still a major cost factor, the brunt of this effort fell on the newly organized manufacturing arm, now headed by a manufacturing manager, whose power and influence grew. The strategies used to stabilize operations were to have serious effects later.

Labor unrest continued during the Depression. With layoffs and speed-ups the order of the day, unions grew in size and power. Although some industries were enlightened and tried to handle labor with wisdom and fairness, economic disaster tended to breed desperate measures. The Amoskeag story ended tragically. Management, deciding it could not justify an investment in the Manchester facility, used the cash flow from operations to fund a trust for the benefit of the owners. Strikes occurred in 1922, 1933, and 1934. The last two were bitter and violent, and in September of 1935 the mill shut down. Liquid assets had been siphoned off to the trust fund, leaving a shattered economy in Manchester, a city of 75,000.

Labor/Management Power Struggles

Union influence varied. On one extreme was Sam Gompers, head of the garment workers union, who actually introduced industrial engineering to the garment trades, lowered costs and saved the needle workers' jobs. At the opposite extreme was the longshoremen's union who allied themselves with gangsters to take over the docks. Abuses and conflict on both sides hardened labor/management relations into adversarial positions.

Automation Extended—The Product Volume Dilemma

Observing the success of mass production and frustrated with the intractable demands of labor, management emphasized increased automation as a solution. Not only was the process automated, but conveyor lines, transfer equipment, and assembly lines were developed to automate material handling. Unfortunately, automation efforts had the negative effect of reducing manufacturing ability to be flexible and accommodate product change. As a result, growing distinctions among types of manufacturing began to be recognized (see Table 2–1).

New products and product styles proliferated, forcing production in the direction of the job shop. But manufacturing management, recognizing the increased costs involved, pushed in the direction of high-volume, minimum flexibility facilities. Striving to meet market needs or stimu-

TABLE 2–1
Three Types of Manufacturing

Type	Product	Characteristics
1. Pure process	Chemicals, steel, wire and cables, liquids (beer, soda), canned goods	Full automation, low labor content in product costs, high-volume output, facilities dedicated to one product
2. High-volume manufacturing	Automobiles, telephones, fasteners, textiles, motors, household fixtures	Automated equipment, partial automated handling, moving assembly lines, most equipment in line, factories dedicated to various models of product
3. Job shops (low-volume)	Capital goods, hand tools, hardware, instruments	Machining centers organized by manufacturing function, not in line, high labor content in product costs, general purpose machinery with significant changeover time, little automation of material handling, large variety of product

late demand, sales and marketing became adversaries of manufacturing, which was looking for a few high-volume products. The conflict sometimes reached the point at which sales and marketing were not allowed in the factory. Complaining about product proliferation, manufacturing tried to freeze schedules. The words "give us long lead times" became the rallying cry of the factory.

Growing Manufacturing Bureaucracy

Overhead continued to increase as material handlers, equipment maintenance and changeover personnel, inspectors, and planning and inventory controllers were added to industrial and manufacturing engineering staffs.

Two unanticipated problems resulted:

1. *Growing Overhead Costs.* Costs of indirect labor (burden) were measured as a percent of direct labor. As direct labor costs were reduced, an increasingly financially oriented management

became alarmed at the rise in percent of burden. Some programs for direct labor reduction were even rejected because overhead percentages increased. In other cases indirect labor was arbitrarily cut. The confusion over burden absorption clouded the issue of cost control throughout this period, even leading to occasional decisions to *add* direct labor to improve burden rates!

2. *Bureaucracy Sets In.* The sheer size and makeup of factory overhead personnel created a conservative, risk-averse bureaucracy. Middle management specialists looked at the production process with tunnel vision and parochial interest. By 1960 the plant staff could best be described as a group of blind men feeling an elephant, each describing it in the limited context of his experience. Not only did the bureaucracy make change difficult; they failed to see the whole process (elephant) and often duplicated, canceled, or offset the efforts of each other. Bright, ambitious activists left for more challenging fields.

World War II Aftermath

With these growing cracks in the manufacturing foundation, why were these the golden years? The reasons lie in the impact of World War II. The prewar depression ensured that the fittest survived. The country entered the 1940s equipped with a lean, mean, and hungry manufacturing community. The conversion of U.S. industry to war production is an incredible story of quick response. Aided by statesmen such as Bob Lovett from the War Department, one of the great manufacturing men of his age, Bill Knudsen from Ford planned, guided, spurred, expedited, and coordinated the conversion. U.S. manufacturers made it work. After the war, the U.S. government recognized that post-World War I reparations had contributed to the 1930s depression and the rise of Hitler. Jack McCloy went to Germany with the Marshall Plan to finance the rebuilding of that devastated economy, and the American Caesar, General Douglas MacArthur, went to Japan to direct its reconstruction. The result was that the United States primed the pump—a growing export market—that sustained the domestic surge of postwar demand. The combined effects of pent-up demand, the export market financed by U.S. loans, and little worldwide competition, generated a huge market for U.S. products. Until 1960 no one could match U.S. costs, but the arrogance of success meant that no one was keeping an eye on Japan and West Germany.

Introduction of Quantitative Methods

Another impact of World War II on manufacturing was the growth of a specialized quantitative problem-solving approach called *operations research* (OR). Operations research had been used to solve a wide variety of war-related problems by teams of multidisciplinary scientists. After their war-time success, some OR scientists turned to manufacturing, attracted by an area full of unresolved problems. This stimulated the development of powerful mathematical techniques and the beginnings of truly *scientific* management applications for manufacturing. From creative chaos emerged an alphabet soup of techniques:

SQC—Statistical Quality Control. Statistics applied first to inspection, then to control processes, and subsequently to problem solving. Used to eliminate defects and their attendant cost.

EOQ—Economic Order Quantity. Mathematical solution of the lot size problem, which trades off setup costs against excess inventory costs.

LP—Linear programming. Complex mathematics that solves for optimum decisions among conflicting costs and restraints.

PERT—Planning, Evaluation, and Review Technique. Project planning to determine the project cost, the critical path, and the shortest time to completion.

MRP—Material Requirements Planning. The explosion of product quantities through a bill of material to determine quantity and dates of required components to meet finished-goods assembly schedules.

LOB—Line of Balance. Line balancing to determine the optimum assignment of tasks to assembly line workers at varying line speeds.

These complex quantitative approaches were sponsored by a growing number of manufacturing consultants and technical societies mirroring the specialization and departmentalization in the manufacturing organization (see Table 2–2).

The specialists pushed their solutions on an industrial population lacking the academic background or training to evaluate and understand them. Top management had neither the experience in manufacturing nor the time to evaluate, select, and sponsor these new techniques. Further-

TABLE 2–2
Some Specialized Technical Societies and Their Functions

Technical Societies	Manufacturing Specialty
ASQC	Inspection and quality control
AIIE	Industrial engineering
SME	Manufacturing engineering
MHI	Material handling
APICS	Product planning, scheduling, and inventory control
NACA	Accounting and cost control
PMA	Purchasing
TIMS	Operations research
AIDS	Mathematicians dedicated to applying quantitative methods to aid decision making

more, they were preoccupied with marketing, product planning, investment strategies, and a growing need to meet regulatory and legal issues.

Impact of the Computer

When the computer arrived on the scene, the young technicians and salesmen of that industry were very much at home with quantitative approaches and immediately saw manufacturing applications and dramatic marketing implications. Most of the new techniques were difficult to handle manually. A hasty marriage of the quantitative specialist with the computer and software industry took place. The technical societies jumped aboard to sell solutions to all of manufacturing's problems.

Top management, mystified by the problems and the bewildering array of acronyms presented as solutions, viewed the computer as an automation device. They rushed toward computerization, partly to "keep up with the Joneses" and partly in the hope that computer automation could replace indirect labor as earlier factory automation had reduced direct labor. The race was on. An unprepared manufacturing bureaucracy tried frantically to upgrade quantitative skills and computer knowledge in the scramble to get smart. There were enough successes to fuel the effort because the problems were real and the techniques were sound when properly applied. However, the consultants and data processing industry generally made more money than the users.

The Final Surge of Demand

Meanwhile, down in the factory, foremen and factory managers focused on the two narrow objectives by which they were measured: Beat the standards and ship the maximum amount of dollars. They "made the numbers" and profits improved. Although labor was still an adversary, its size had dropped to between 10 percent and 20 percent of cost of sales for job shops and to under 10 percent in high-volume manufacturing. Flexibility was a problem and factory overhead was climbing, but GNP grew and productivity increased almost every year. Who said there were cracks in the foundation?

1960–1980, THE DECLINE OF THE AMERICAN FACTORY

The Turning of the Tide

What then finally went wrong? Perhaps the best example is the automotive industry. A remarkable full-scale account of this period as it affected Detroit and the automotive industry can be found in David Halberstam's *The Reckoning*. In 1960, General Motors held a substantial cost advantage over its rivals. Economies of scale, good engineering, and styling to meet customers' whims had given it the lead. Ford, the early leader based on its strength in manufacturing innovation, had slipped badly into a poor second. GM did not dare drop prices to reflect its cost advantage. Fear of government action under the Sherman Antitrust Laws required a concerted effort to keep at least two other viable competitors in business. The law had the unintended effect of actually reducing competition.

1973 saw the beginning of a general slowdown in U.S. industrial growth. Supply had finally caught up with demand. The investment in reconstruction of Japan and Western Europe, particularly West Germany, had brought new factories online. These countries began to import less as revived industries prospered in domestic markets, supported by the hidden import duty of a high U.S. dollar. "Stagflation" appeared in the United States (aggravated by the OPEC crises and increased oil prices). Productivity leveled off. Because U.S. management had been measured on earnings per share and return on investment, management did not want to look worse by investing in plant and equipment. As volume and profits sagged, the desire to look good in the short term was achieved by running depreciated and increasingly obsolescent equipment—shades of Amoskeag!

The Japanese Challenge

The 1970s were bad enough, but worse was yet to come. Japan started to ship attractive cars in increasing quantity to the United States at $1,000 lower cost than the best U.S. producers. Toyota is generally given credit for the low-cost manufacturing edge. Japan had accepted MacArthur's admonition that their future lay in economic competition as opposed to disastrous military adventure. First they rebuilt their industrial base, which they then matched with an educational program particularly strong in mathematics and engineering. They accepted the old criticism of their prewar consumer products as poor quality imitations, and set out to do better. Large teams of industrial managers visited the States in the 1960s. They learned all they could about American manufacturing. Moreover, they learned with a critical eye and with two extremely wise philosophies: Keep it simple, and deal with the whole, not just the parts.

Like Francis Cabot Lowell 100 years earlier, they put together what they had seen in an innovative and brilliant adaptation. Nearly all the elements of the Toyota concept were gleaned from the United States. The breakthrough came from the way the pieces were combined in a "holistic" approach. Toyota's cars were cheaper, were better designed, were less costly to maintain, had fewer defects, and (especially valuable in the OPEC crisis) had lower gas consumption. Table 2–3 shows the cost

TABLE 2–3
1977 Hertz Incidence of Repairs Study[a]

Model	Repairs/100 Vehicles
Ford	326
Chevrolet	425
Pinto	306
Toyota	55

[a] First 12,000 miles.

advantages as seen through the eyes of a U.S. car rental agency. Figure 2–2 shows a productivity comparison between a U.S. and a Japanese engine plant.

Detroit's answer was a typical response to any catastrophe: first denial, then rage, followed by recognition of the problem, and finally a belated response.

Admittedly Japan had lower labor costs and a much newer manufacturing infrastructure. But their impact was far more profound because their strategy and system of manufacturing produced a quantum jump in product-cost reduction. Because of some of its more obvious aspects, the Toyota system was called just-in-time (JIT).

The Just-In-Time Principles

JIT principles in hindsight seem obvious, but they had not been part of the U.S. philosophy (for the following discussion of JIT, the writer is indebted to Ed Hay of Rath & Strong, Inc., Lexington, Massachusetts):

1. Eliminate waste: Waste is anything that does not add value to the product.
2. Henry Ford was correct: The most efficient way to manufacture is to be able to make and move one unit at a time—not just in assembly, but throughout the whole process.
3. If you don't need it now, don't make it now: Produce each day only what is sold each day.
4. Repeal Murphy's law: Instead of living with problems, initiate a process of continuous problem solving.

FIGURE 2–2
Focused Factory

	Toyota	Ford
Engines/Day/Employee	9	2
Square Feet	300,000	900,000

Toyota's productivity level is 4.5 times Ford's, with 2/3 to 1/2 lower capital investment.

JIT Approach to Labor

JIT looked at labor with fresh eyes. If direct labor represented less than 20 percent of the cost, obviously cost improvement and productivity gains must lie elsewhere. Toyota backed off the philosophy of deskilling and speeding up labor. In contrast, it involved the direct labor work force in some of the work usually done by indirect labor or the manufacturing staff. These activities included quality control, maintenance, clean up, and problem solving. Time and training required to update labor skills was provided during the work week in sessions called *quality circles*.

JIT Approach and Inventory

Inventory was looked upon as waste, not as an asset. For years U. S. foremen had used inventory (making more than required at the moment) to increase output (long runs), bridge long changeover times, level out demand variation, and cover breakdowns or quality losses. JIT recognized that inventory covered other problems. The Japanese reduced inventory until production was interrupted (called *stressing the system*); then they solved, for example, the quality control problem before continuing production and reducing inventory further to uncover the next problem. In the United States, stopping production was a mortal sin. Any inventory reduction that jeopardized output only proved that inventory was necessary—not just in time, but just in case. JIT looked at the size of inventory as a measure of how much progress was being made in problem solving. For that reason JIT has often been called *Zero Inventory*, a definition that is far too narrow.

JIT Approach to Quality

The concept of eliminating waste attacked another Achilles' heel of U.S. industry. Quality and reliability problems had been reduced in America by applying the methods of giants in the domestic quality control field: Juran, Shainin, Seder, Fiegenbaum, and others. They were frustrated at being able to stimulate only a portion of the potential progress using their methodology for making it right the first time. U.S. management assumed quality improvement was costly, and went only as far as competition required. The quality control advocates successfully turned to military and space hardware manufacturing, where there was no room for defects. In frustration, one of them, Deming, went to Japan; he became almost a deity. The Japanese recognized all the costs of poor quality (inspection, lost labor and material, inventory, salvage, etc.) and vigorously attacked the causes. Production costs dropped and customer satisfaction increased.

JIT Approach to Product Flow and Flexibility

U.S. factories had developed serious layout and run-size problems. JIT rearranged the factory, improved the flow, and shortened supply lines.

U.S. factories were divided into job shops (equipment clustered in functional groups—all milling machines together, all grinders together) or high-volume manufacturing (equipment in line according to product operation sequence), depending on volume and product variety. The auto industry's assumption was that demand would permit high-volume manufacturing. However, model changeover, variety, and options had demonstrated that massive, fully automated production lines were too rigid and therefore too expensive to meet market changes. Some factories became a mixture of job shop and high volume with less than optimum flow.

The JIT approach became one of flexible machining centers (job-shop concept) that could be moved as product changes warranted (product-line concept), with flexible labor capable of running any and all of the machines required. Labor was balanced and machine speeds adjusted to run at the rate the schedule required, not maximum machine speed. In the United States, a foreman would generate an "unfavorable variance" if a machine was slowed down (underutilized). The accounting and performance measurement systems guaranteed poor flow as each machine had to run at maximum speed. The result was considerable interruption and inventory buildup as operations were started and stopped.

To achieve smooth flow and still accommodate product variety under the JIT concept, machine changeovers were analyzed just as the early industrial engineers had studied direct labor operations. Changeover time was reduced until there was no penalty for running a single piece. The flow was no longer interrupted when the same machines were used to make different items. The result was a *flexible factory* that could run at any speed and tolerate product changes paced to customer demand.

The difference between job-shop and high-volume manufacturing became less distinct as factories sought the benefits of both: product variety with assembly-line flow, that is, flexibility and low cost. Such a factory produced only what was needed rather than building up inventory at high speed and then shutting down to use it up.

JIT Production Objectives

JIT made the schedule the goal. Labor stopped only when the schedule was complete, regardless of the time of day. If schedule completion took place before the end of shift, workers stopped producing product and went into training, problem solving, clean up, or other nonproduction job activities.

As the flow improved and products moved quickly, dramatic results occurred. Work-in-process inventory dried up, causing lead time to shrink, making manufacturing far more responsive to customer changes. Simpler systems were possible because there was less semifinished material on the floor at any one time.

JIT manufacturing strategy is best summarized by an old industrial engineering principle: Improve the method before setting the rate. Only after the factory is properly focused on the optimum arrangement and product flow would automation be applied, whether in terms of equipment (robotics, transfer equipment) or computerization (computer-aided manufacturing—CAM). The United States would occasionally automate a mess by developing expensive equipment and systems to deal with an environment that had not been optimized.

The JIT measure of performance was final product output versus schedule—more being as bad as less. Total manufacturing cost per unit produced, rather than direct labor variance from standard became the key number in evaluating manufacturing productivity.

Managing the JIT System

In well-managed Japanese industries, top management served an apprenticeship in all departments, including manufacturing. At the corporate planning level, they married product planning to manufacturing planning. Significant changes in product design would generate revisions to the factory. The rebuilt Japanese manufacturing base would be constantly revamped and upgraded with substantial investment. Japan avoided the cancer of obsolescence leading to decline demonstrated by the Amoskeag story and, more recently, the U.S. steel industry. Japanese managers took a longer-range view.

In the manufacturing organization, there was one technical arm—

production engineering. These staff-support technicians were taught manufacturing not as a group of differing disciplines and techniques but as an application of a common body of knowledge integrating industrial engineering, manufacturing engineering, quality engineering, planning, scheduling, and factory accounting principles. The manufacturing organization was flat: factory manager to foremen supported by a relatively small but well-trained production engineering staff.

Japanese labor management requires a very high level of worker loyalty, discipline, and dedication, which may well change with time and affluence. Several JIT principles did break down the adversarial and dehumanizing elements of the U.S. approach.

1. Labor was cross-trained to be flexible enough to run a large variety of equipment.
2. Labor was not tied to individual output goals (the standard incentive plan) but to meeting schedule.
3. Direct labor could stop the process to force problem solving rather than running defects for subsequent repair and rework.
4. Downtime was utilized to perform maintenance, housekeeping, problem solving (in quality circles), and cross-training.

In short, direct labor approached a salaried-employee concept, although the levels of labor were adjusted periodically to match the rate of customer demand. Finally, burden rates were divorced from labor (labor-based accounting) and applied to the process (process accounting). This eliminated variance control and burden absorption as a misleading and unreliable basis for decision making.

JIT Cost Performance

The average cost improvement from JIT, primarily resulting from a continual problem solving effort, has been about a 40 percent reduction in manufacturing costs. The bulk of the savings came in overhead costs: inspection, changeover time, material handling, scrap, rework, storage, accounting and staff support. A significant saving comes from smaller factories per unit of output. As in the United States, direct labor reduction is a function of automation, but only when the factory is ready for it. The lowest cost to achieve growth through increased market share is the prime goal, with earnings per share and return on investment as measures of success rather than ends in themselves.

Domestic Organizational Developments

During the 1960 to 1980 period of decline, there were some significant developments in U.S. management. An output of the sixties was the "human potential" movement. Industrial psychologists had been around a long time, as the Hawthorne experiments in the 1930s illustrated, but behaviorists such as B. F. Skinner at Harvard began to look toward a broader application of their theories in industry. A growing number of behavioral scientists suggested that companies needed to utilize the potential in the work force and lower-level management. Large and complex organizations, full of technical specialists, had evolved into a collection of separate departmental functions. Each knew more and more about less and less, and were managed by a small top management group who seemed to know less and less about more and more, at least as viewed from the manufacturing perspective.

Coalescing around the acronym OD (organization dynamics), behaviorists challenged the traditional hierarchical structure with a bottom-up approach, a concept requiring extensive behavior modification. The OD advocates also emphasized the importance of the *process* of organizational interplay rather than the pure *content* of the specialities (engineering, scheduling, accounting, etc.). While the Japanese had adopted a consensus style of bottom-up management, the OD fraternity went further. They addressed not only low-level input and consensus, but suggested revamping the authoritarian approach to one that would be more effective in managing change and avoiding higher-level management isolation. David Halberstam's *The Reckoning* indicates dramatically the severe problems arising from that isolation.

As of 1980, a conflict between OD and the old military-high-command approach had developed. OD became known as the "soft science," the implication by the traditionalists being that tough decisions and strategies would suffer. The "soft" scientists and the combat-weary generals with their engineering-trained "hard" scientists mixed like oil and water.

Business schools also appeared to be deeply divided on this score: Harvard and Stanford advocated the traditional top-down, aggressive, financially-oriented decision making; Yale was the home of the behavioral, consensus concept, more sensitive to human potential. MIT's Sloan School was the prime advocate of system, content, and technology, as well as the need for technically trained management. However, all of these graduate programs are now in various stages of transition, and perhaps the ultimate revision will contain the best elements from each school. At the very least, it appears that the labor/management problem is finally moving out of an adversarial stand-off, and the "military school of industrial command" will be significantly modified to meet and manage change.

THE END OF THE 20TH CENTURY— CHALLENGES AND RESPONSE

As of 1989, there is evidence that many U.S. manufacturers have made significant strides, assisted by the falling dollar and low oil prices. Manufacturing is still a mystery to a large segment of the financial, political, and academic communities. The prestigious graduate schools of business are just beginning to look at manufacturing as something more than a low-level effort that happens between order booking and shipment. But Japan and West Germany have wielded the two-by-four that got U.S. management's attention. At the end of the Reagan administration, significant cost reduction (often called *restructuring*) using some of the JIT concepts has dramatically improved manufacturing costs. Much to the surprise of economists and Wall Street, profits and productivity of well-managed U.S. companies are improving. There is more to be done, however, and some industries may never recover.

The Challenge

This country has used adversity (the 1929 depression and the war years) to strengthen the survivors and reach extraordinary new levels of competence. Overseas competition may well be another trial by fire. It may force key revisions to manufacturing strategy and reestablish a low-cost base, balancing the skills at which U.S. manufacturing still excels: innovative and entrepreneurial attitudes, engineering ability, marketing acumen, financial sophistication, and legal or regulatorial ability. This review of manufacturing history has tried to suggest strategic considerations and revisions to restore our competitive edge in the world and to provide a framework for subsequent chapters in this book. Six recommendations to U.S. management are offered:

1. Develop a coherent body of knowledge for both academic and in-house educational programs as a means to train more competent manufacturing personnel. Financial and engineering staffs learn their trade with the aid and support of excellent academic programs, but not manufacturing. An increasingly complex world requires more than warm bodies to fill manufacturing management ranks.

The successful model of airline pilots' and computer programmers' training suggests that a degreed graduate program is not the only solution, and the body of knowledge from which to teach still needs drastic revisions and general agreement on content.*

2. At all levels of academia and industrial management, recognize the importance of attracting (and maintaining) the brightest and best minds into manufacturing. Today, manufacturing is not a career path to the top. We have a vicious cycle to break: Good students are not attracted to manufacturing courses, and good teachers are not attracted to teach them. What MBA is going to go to Peoria, Illinois for $35,000 a year to work in a factory, when a $100,000 consulting, arbitrage, or merger and acquisition job is available in New York, Boston, or San Francisco?

3. Revamp traditional cost accounting to reflect how costs really act. Methodology exists in the realms of *direct costing* as an improved method of dealing with fixed and variable costs and *process costing,* to move away from labor-based overhead accounting. Managing to the wrong objectives is hardly the intent of management by objectives. Both the NACA, the CPA firms, and accounting courses must be involved. Productivity must be redefined, and manufacturing goals and measurements must be redirected to a variety of goals (quality, delivery, share of the market, return on investment, earnings per share). Goals and performance measurements must support and reward revised manufacturing strategy.

4. At the highest levels of management, reconsider the financial bias toward short-term returns that has inhibited investment in technology at critical times in industry. Francis Cabot Lowell in 1810 and Japanese industrialists in 1970 recognized that industry is *technology-driven.* U.S.

---

*Editor's Note: The MIT Leaders in Manufacturing program, a new innovative dual degree at the Sloan School, combines MBA and Masters engineering work to produce a new kind of manufacturing leader. See Chapter 1.

industry has been increasingly *financially* driven since World War II. Success is impossible without sound marketing and financial skills, but the best marketing organization with the largest financial resources can only delay death from technological obsolescence of product or plant. Can we resist the feeding frenzy of the 1980s financial markets and fund a continual investment program to regain and maintain technological leadership in our manufacturing base?

5. With the exception of pure process industries (chemicals and oils, for example), every company should thoroughly understand what is behind the acronym JIT and develop a tactical plan to move toward or improve on that strategy. Toyota took ten years to pioneer the concept, making many mistakes along the way. With large bureaucracies to challenge, to convince, and to restructure, the task for American manufacturers will take time as well as committed top management. Fortunately, savings accrue from the first years, but the effort must continue over time. Zero inventory, zero direct labor, zero defects may never be attainable but are important targets. U.S. management must replace fad management (jumping from SQC to CAD, CAM, MRP, OR, OD, etc.) with a program of continual problem solving under a sound production strategy.

6. Develop an organization and set of informing mission statements that:

- Manage change. The exponential growth of technology suggests that changes will continue to accelerate. Flexible factories, people, and systems are required.
- Upgrade skills at all levels. We need to be smarter as automation and computers perform manual clerical work. We need prepared workers to plan, make decisions, and solve problems.
- Manage specialization. The tools and systems of manufacturing incorporate machine design, applied statistics, computer systems, materials science, behavioral science, and so on. Manufacturing managers need to understand the application of and effects of using resources that necessarily evolve from specialists. The broader liberal arts curriculum answers the need for people who can deal with specialists within the context of understanding the total manufacturing environment. Manufacturing is a man/machine interface better managed by generalists than specialists — a lesson that the Japanese have demonstrated.

TABLE 2–4
Quasar: U.S. Operations

	Prior Management	JIT Management (2 years later)
Direct Labor Employees	1,000	1,000[a]
Indirect Labor Employees	600	300
	1,600	1,300
Daily Production	1,000 Units	2,000 Units
Assembly Repairs	130%	6%
Annual Warranty Costs	$16M	$2M[b]

[a] Same employees.
[b] 2 × units or 16:1 improvement.

Is all this possible? Table 2–4 shows the performance of a TV assembly facility located in Franklin Park, Chicago, before and after it was acquired by a Japanese company using the JIT system. The facility is run by the same work force and is still located in a high labor cost Northern urban environment. The only change is a management team with a different strategy and philosophy.

The opportunity exists. The need is critical. Quasar is not an isolated example. The next period for U.S. manufacturing has begun. Our world leadership in manufacturing can be regained, but it will take a change in management's perception, interest, and understanding of manufacturing as a profession. It will also take a manufacturing staff of fewer, brighter, and better professionals with the talent and broader educational background to pull it off. This text suggests that we already have the tools and the know-how.

BIBLIOGRAPHY

Clark, Kim; Hayes, Robert; and Lorenz, Christopher. *The Uneasy Alliance.* Cambridge MA: Harvard Business School Press, 1985. See chapter by Wickham Skinner, " The Taming of Lions: How Manufacturing Leadership Evolved 1780-1984," pp. 63–110

Dalzell, Robert F., Jr. *The Enterprising Elite—the Boston Associates and the World They Made.* Cambridge, MA: Harvard University Press, 1987.

Halberstam, David. *The Reckoning*. NY: William Morrow and Company, Inc., 1986.

Hareven, Tamara, & Langenbach, Randolph. *Amoskeag, Life and Work in an American Factory City*. NY: Pantheon Books, 1987.

Hay, Edward J., Jr. *Just-In-Time Breakthrough: Implementing the New Manufacturing Basics*, NY: John Wiley & Sons, 1988.

Skinner, Wickham. "The Productivity Paradox." *Harvard Business Review*, July–August 1986, pp. 55–59.

PART 2

THE STRATEGIC MANUFACTURING PLANNING PROCESS

Strategies and tactics. Strategy refers to competitive approaches toward winning a market niche; being best in quality or lower in price, for example, or offering the most features or the newest idea. Tactics are the methods of supporting and executing the chosen strategy. In seeking to be the market innovator, for example, appropriate tactics would include large expenditures for research and development and a very flexible and fast new product introduction capability.

In Chapter 3, four managers from Hewlett-Packard fully describe the manufacturing strategy process, including industry and company analysis. A third component of the planning process currently receiving more press in the United States is the element of global competitive requirements, covered by Linda Sprague in Chapter 4. Finally, the question of how much planning is enough is answered in Chapter 5 by Jeanne Leidtka.

CHAPTER 3

USING MANUFACTURING AS A COMPETITIVE WEAPON: THE DEVELOPMENT OF A MANUFACTURING STRATEGY

Sara L. Beckman
William A. Boller
Stephen A. Hamilton
John W. Monroe

Editor's note

In this chapter, four authors explain the methodology that Hewlett-Packard has developed—a five-step strategic manufacturing planning process that can help a company reestablish manufacturing's position as a critical contributor to the success of the company's business strategy. They show how this planning process is particularly effective in helping companies compete when measured by critical success factors, like quality, features, price, and availability. The process, which combines academic theory and real world applications, works for HP's customers as well.

INTRODUCTION

In 1958, John Kenneth Galbraith wrote that the United States had solved the production problem. Nothing has proved to be further from the truth. As management attention was diverted from improving manufacturing capability to problems of mass distribution, advertising, and marketing, U.S. manufacturing capabilities began their decline. Meanwhile, some of the more successful international competitors identified a "new" secret weapon: manufacturing. As numerous recent studies report, U.S. manufacturing productivity has declined severely, and much of the product we now consume is manufactured in other countries.

Many reasons are cited for the U.S. manufacturing crisis. Some claim that lower labor rates in other countries are to blame, and they chase cheap labor around the world in an attempt to manage costs. Others believe that foreign competition is not as burdened by government regulation as are U.S. manufacturers; they seek changes in U.S. policy and tax structures. Still others attribute the loss of a competitive edge to a general deterioration of the American work ethic. They try to understand the cultural contribution of the competitors' environments and argue the feasibility of transferring cultural characteristics to the U.S. worker.

While all of these problems, and others, are likely contributors to the U.S. competitiveness dilemma, no single one is the sole cause. The beginnings of the U.S. decline appear to be rooted in much more distant history. Wickham Skinner traces the origin of the problem to the turn of the century, when top management in America stopped making manufacturing issues (process technology development, equipment innovation, production architecture) principal concerns in their strategic agendas. The recent President's Commission report on global competition corroborates Skinner's observation. It claims that there has been a general failure by American industry to devote enough attention to process technology and to manufacturing as a whole.

Clearly, there is a significant need for U.S. manufacturers to recognize the critical role that manufacturing plays in the accomplishment of business objectives and to integrate the manufacturing function as a partner in strategy development and execution.

AN APPROACH TO INTEGRATING MANUFACTURING

At Hewlett-Packard (HP), we have developed a strategic manufacturing planning process that we believe can help your company re-establish manufacturing's position as a critical contributor to the success of the company's business strategy. It combines many of the techniques proposed by the academic community, tools from real world manufacturing, and the collective U.S. manufacturing experience to create a process that is understandable and that works. It works not only for us, but has been proven by some of our major customers from aerospace, food processing, telecommunications, and other industries.

Our five-step approach to manufacturing strategy development is as follows:

1. Start with the business strategy. More specifically, understand why customers will prefer your product or service to the competitors'.
2. Create a manufacturing strategy that is *linked* to the business strategy. In other words, specify manufacturing's contribution to making customers chose your product or service instead of your competitor's.
3. Identify manufacturing tactics to execute the strategy. This requires understanding how to manage and control the people, processes, materials, and information needed to deliver products or services in a way that meets the objectives of the strategy.
4. Organize for manufacturing success. Organization design, including structure and performance measurement, must match strategic needs or success will be limited.
5. Measure the results and initiate further change. Strategies must be continually altered to meet the needs of a constantly changing environment. Feedback loops are critical to the continuous improvement process.

This process must begin at the top. The highest levels of management must commit the corporation to a strategic direction that is clearly defined and well-communicated throughout, and conduct periodic reviews to ensure that its direction is being followed.

Here in more detail is a description of the strategic manufacturing planning process.

START WITH THE BUSINESS STRATEGY

What is a business strategy? Unfortunately, *strategy* is both an underdefined and overused word. In the most basic sense, a definition of strategy is the *science* and *art* of military *command* exercised to meet the enemy in combat under *advantageous conditions*. Current management literature tailors this definition to the business environment. Hayes and Wheelwright suggest that business strategy ". . . specifies the scope of that business in a way that links the strategy of the business to that of the corporation as a whole, and (describes) the basis on which that business unit will achieve and maintain a competitive advantage."[1]

Most authors agree that a business strategy must:

1. Describe the methods of competition (e.g., occupy a specific niche of the marketplace that is not currently occupied by other major competitors—Niche A).
2. Define the contributions of each product and function to the goals of the business (e.g., low-end product offerings will keep low-end competitors from moving into Niche A).
3. Allocate resources among products and functions (e.g., manufacturing will receive X dollars to develop flexible manufacturing capabilities for building custom products for customers in Niche A). Competition is the focus of the business strategy. The strategic statement must describe both the customer-driven goals of the business and the allocation of resources to achieve those goals, and it must be based on facts about the competition.

Development of a business strategy, a necessary prerequisite to generation of a manufacturing strategy, is a difficult and iterative process. It depends on input from many different sources. External sources (customers, competitors, and the economic environment) should drive some objectives, while an assessment of internal strengths and weaknesses should drive others. General Foods, for example, has been able to gain a significant competitive advantage in the dessert foods marketplace by employing its proprietary process for making gelatin products. In fact, antitrust suits have been brought against General Foods because they have such a large proportion of the market. By understanding the desires of the customer and by using its process capabilities wisely, General Foods has resoundingly beaten its competitors in this marketplace.

Critical Success Factors

Ultimately, a business strategy must begin with the customer. Why will customers buy your products or services rather than those of your competitors? We defined a short list of possible answers. Customers buy products or services based on their perception of one or more of the following characteristics:

* Low product or service cost.
* High product or service quality.
* Prompt product or service availability.
* Distinguishing product or service features.

Definition of your desired position in each of these characteristics vis-a-vis your primary competitors virtually defines your business strategy.

Business Segmentation

Difficulties may arise during the process of identifying critical success factors, however, if your company is actually in more than one business. In this case, you must include a business segmentation exercise as a part of your strategic planning process.

A business segment is a set of products or services and customers that shares a distinct set of economics. The opportunities to obtain competitive advantage through manufacturing may differ significantly by segment. It is therefore critical that a proper segmentation be made early in the planning process. While there are many ways to segment a business, the only meaningful one is based on the real needs of the customer and the cost structure required to meet those needs.

An example might make this a little more clear. At HP, we are in many different businesses. Customers for our test and measurement instruments (e.g., oscilloscopes) require a wide variety of product technology with lasting value—which implies broad product mix, inclusion of current technology, and support for the old technology. A critical success factor in this market is providing features. The minicomputer (e.g., HP 3000) marketplace, on the other hand, requires much less hardware product variety. (In fact, product standardization is becoming an issue for these customers.) Market growth is relatively quick, and new computing technologies are being developed rapidly. Two critical success factors in this business are quality and features. Finally, customers for the personal computers (e.g., Vectra) and terminals virtually dictate high volume and a high degree of product standardization. In this marketplace, price and availability are the most critical success factors. Figure 3–1 summarizes the three different businesses.

FIGURE 3–1
Important Product/Market Characteristics

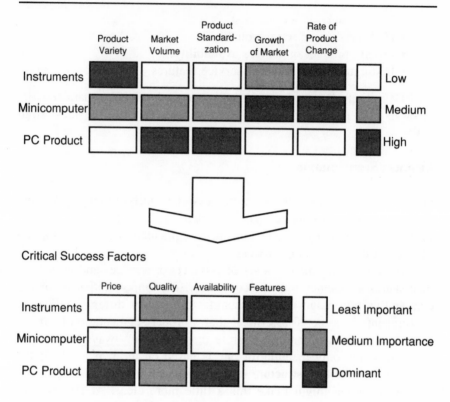

Critical Success Factors

Note that while each business emphasizes different critical success factors, it does not ignore the others. All businesses will pay some attention to cost, quality, availability, and features. Focus on the primary success factors, however, is critical.

Business segments do not remain static. They will change over time as customer requirements, internal capabilities, competitors' strategies, available technology, and the economic environment change. Further, they are likely to change as products or services move through the various stages of their life cycles. Thus, constant monitoring and re-evaluation of the true business segmentation and the associated manufacturing missions are essential to long-term success.

A coherent, well-focused business strategy allows all of the functions in a manufacturing business to work together with a shared vision of the organization's purpose and direction. It provides the guidance for all decision making, resource allocation, and performance measurement activities in the organization, as well as a common basis for communication throughout. It is the critical first step in the strategic manufacturing process.

CREATE A *LINKED* MANUFACTURING STRATEGY

Manufacturing, marketing, R&D, finance, and human resources must all create strategies that collectively and synergistically comprise the overall strategy for the business. These strategies are referred to as functional strategies. Each may have a slightly different emphasis on achievement of one or more of the objectives of the business strategy.

For example, in its management of the customer interface, marketing may be primarily responsible for the availability objective. R&D may take more responsibility for the features of the product in the product development process. R&D should also take responsibility for a large portion of the quality and cost objectives, as they are both greatly affected during product design. However, manufacturing may retain significant responsibility for cost and quality as well for the management of both new product introduction and ongoing production. We are chiefly concerned here with manufacturing's role.

Just as there are numerous definitions of business strategy, there are several definitions of manufacturing strategy. Hayes and Wheelwright define a manufacturing strategy as "a pattern of decisions actually made, and the degree to which that pattern supports the business strategy. . . . It is the collective pattern of these decisions that determines the strategic capabilities of a manufacturing operation."[2] Roger Schmenner puts it more simply when he defines a manufacturing strategy as "a plan that describes the way to produce and distribute the product."[3]

Most authors agree on a few common characteristics of a manufacturing strategy.

- First, the manufacturing strategy must support the business strategy by focusing the manufacturing activity on a small set of objectives dictated by customer need.
- Second, it should describe allocation of resources within the manufacturing function in a way that allows achievement of the manufacturing objectives.

• Third, it will be reflected in the patterns of actual decisions (e.g., capacity expansion, automation) made by the manufacturing function. If the decisions made over time by the organization are not consistent with the stated strategy (and many times they aren't), then the stated strategy is displayed by the enacted strategy.

A key aspect of *linking* the manufacturing strategy to the business strategy is to have manufacturing team members determine the contribution that manufacturing can make in each of the areas of customer need: cost, quality, availability, and features. Providing an advantage in the *cost* category implies being able to produce products or services for the lowest total cost, where total cost includes cost of ownership throughout the product or service life. Advantages in *quality* must be obtained from some conformance to or betterment of customer requirements for products or services. *Availability* measures the ability to deliver the product when and where desired, as well as the ability to respond to changes in market demand and opportunities. Manufacturing makes a contribution to an availability objective by being flexible and/or responsive. Finally, manufacturing can contribute product or service *features* whenever its process allows inclusion of unique attributes in the product design or service. This requires that manufacturing be innovative in its development of new process technologies. Integrated circuit fabrication facilities, for example, often allow inclusion of more capability in an integrated circuit, thus encouraging use of new attributes in a product design.

Following through with our HP example (see Figure 3–2), we see that manufacturing's strategic emphasis in the instrument business is on flexibility/responsiveness and innovation/technology, which allows it to respond with the features required by its customers. Minicomputer manufacturing's emphasis is on quality and flexibility/responsiveness, which allows it to respond to requirements for both quality and product features. Finally, personal computer manufacturing must focus on cost and responsiveness to meet its customer needs for low-cost and off-the-shelf availability. Again, note that while each manufacturing strategy differs in its primary emphasis, all elements have some role to play in the strategy.

It is interesting that American manufacturing has traditionally focused on cost in its effort to maximize machine and human utilization in the manufacturing process. The Japanese, on the other hand, have had

FIGURE 3–2
Critical Success Factors

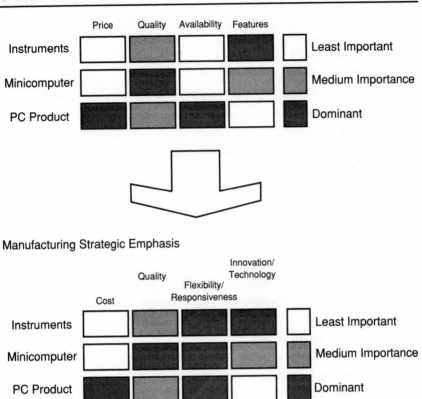

a nearly unilateral obsession with quality. Both may be right; it depends on what the customer wants. Manufacturing can only be supportive of the business strategy when the strategic objectives are made clear and the contribution of the manufacturing function is well-understood and described.

Strategic Decision-Making Categories

Once a manufacturing organization understands its points of strategic emphasis (cost, quality, flexibility/responsiveness, or innovation/technology), it has the basis for making further strategic decisions. These decisions fall into a number of categories, which may include the following:

1. Capacity/Facilities
   How many sites?
   How large?
   Where located?
   How focused (e.g., on products, markets, processes)?

2. Work Force/Organization
   Skill sets required?
   How measured and compensated?
   What composition?

3. Information Management/Systems
   How structured?
   Who owns?
   Degree of automation?

4. Vertical Integration/Sourcing
   How vertically integrated (forward or backward)?
   How many suppliers?
   What types of supplier relationships?

5. Process Technology
   What types?
   How automated?
   Make or buy?

6. Quality
   What is it?
   How measured?
   Who's responsible?

This list is but one of many. Each company will want to select its own list of characteristics most relevant to the execution of its business and manufacturing strategies. The result of this exercise may be a matrix like the one shown in Figure 3–3.

Consultants and other outside observers of manufacturing companies often examine the patterns of decisions made by the companies in categories such as the ones shown in Figure 3–3 to determine the de facto manufacturing strategy of the firm. For example, a company that continues to invest in hard automation dedicated to the manu-

63

FIGURE 3–3
Manufacturer's Strategic Decision Making Matrix

	Cost	Quality	Availability	Features
Capacity/ Facilities	+	+	++	
Workforce/ Organization	+++	+++	++	+
Information Management	++	++	+++	
Vertical Integration/ Sourcing	+++	+++	++	+++
Process Technology	++	++		
Quality	+++	+++	+++	+++

+   Least important
++   Medium importance
+++   Most important

facture of a single product is unlikely to be pursuing a strategy of flexibility to accommodate rapid product change. If they claim to be, they need to be told that their pattern of decision making is not supportive of their stated strategic direction.

Some Tools for Manufacturing Strategy Development

Most business strategies focus primarily on product- and market-related characteristics rather than on manufacturing- or process-related items. A couple of tools that help translate product-focused business strategy to process-focused manufacturing strategy are useful.

Hayes and Wheelwright[4] formalized the concept of the product/process life-cycle matrix displayed in simplified form in Figure 3–4. In this matrix, products are characterized as falling into one of three categories, often due to their position along the product life cycle. Products

FIGURE 3–4
Product/Process Life-Cycle Matrix

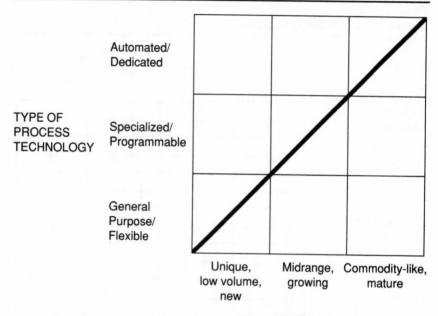

that are unique, customized, or sold in very low volumes comprise the first category. Products often fall into this category when they are first introduced. On the other end of the spectrum are highly standardized, commodity-like products that are likely sold in very large volumes. These are typically considered to be mature products that are at the peak of their product cycle. In the middle are products that are relatively stable and standardized and that are sold in medium volumes, usually into a growing marketplace.

Intuitively, one can define the characteristics of the process technology required for each product category. Unique, one-of-a-kind products require manufacturing flexibility. A job-shop or general-purpose process environment best meets this requirement. For midrange products requiring superior quality and room for innovation, a specialized or programmable technology is appropriate. The mature, standardized product must be consistent and must be produced at the lowest possible cost. This is achieved with highly automated and dedicated process technology. Thus, we define the types of process technology along the other dimension of our matrix.

In general, if a production process is properly mapped to the requirements of the product it is to build, the facility should be found on the diagonal of this matrix. Dedicated automation will rarely be found in a factory intended to build many different, unique products. Similarly, programmable automation is likely to be too costly for a facility building commodity-like, standardized products.

Understanding the position of your company's products on this matrix should provide some useful insights. For example, do all products require the same type of technology? If not, should some products be redefined? Or should the process technologies currently employed be changed? If more than one technology is employed, should separate plants be established, or can they share one facility? Where are your competitors positioned? What position should be assumed to obtain an advantage over them? Do planned new product introductions fit with existing process technology? What types of capital equipment should be purchased to meet future needs? Overall, the matrix provides a useful thinking tool and framework for the development of a manufacturing strategy.

A recent article on the development of the Korean microwave manufacturing capability exemplifies movement on this matrix.[5] When the Koreans first began to develop their version of the microwave oven, their production process was almost entirely manual. As they developed ovens that they were able to sell in volume to a world market, however, they were able to purchase automated equipment for their manufacture. Further, they were able to move from doing primarily assembly types of activities to actually fabricating some of the critical components of the ovens themselves in dedicated, automated facilities. Over time they moved from the lower-lefthand to the upper-righthand quadrant of the matrix.

Another concept useful to the creation of manufacturing strategies is *factory focus*, popularized by Wickham Skinner in 1974.[6] A simplified version of the concept is depicted in Figure 3–5. An operation with a small number of process technologies is called *process focused*. Steel producers are commonly thought of as being process focused because they use a common set of processes to produce a small number of fairly homogeneous end products. An operation manufacturing similar products, possibly using several different process technologies, is called

FIGURE 3–5
Simplified Version of Types of Factory Focus

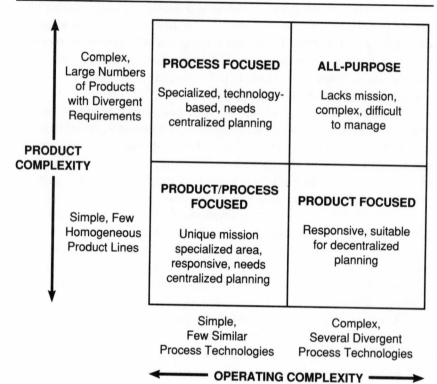

product focused. Most electronic equipment manufacturers have product focused facilities. They produce televisions, computers, or video recorders using whatever collection of processes is required to do so (e.g., printed circuit board fabrication and assembly, mechanical assembly, final product assembly and test). When a facility produces a large number of products and uses many different technologies it is said to be an all-purpose facility.

A company with a few homogeneous product lines can have a simple infrastructure for managing its processes, whereas a company with a large number of products with divergent requirements usually has a more complex production process. Product-focused facilities are often believed to be most desirable; however, process-focused facilities can be equally effective in some circumstances.

All-purpose facilities are generally to be avoided. One can imagine the difficulty in achieving any particular competitive advantage from a facility that is attempting to be all things to all customers. Often, after years of rapid growth and many new product introductions, a facility will lose focus and become all-purpose. Companies that find themselves with such facilities should seriously consider change. Can the facility be separated into a number of smaller product- or process-focused facilities? Should the facilities be separated physically or organizationally? Can the problem be solved by creating a plant within the plant? Having a focused factory is considered critical to long-term success.

Both of these concepts—product/process mapping and factory focus—allow the user to identify alternative manufacturing strategies and to map these alternatives against the requirements of the business strategy and, ultimately, of the customer. The manufacturing strategy must describe the contribution that manufacturing makes to the cost, quality, availability, and features objectives of the business and must provide direction for achieving that contribution.

IDENTIFYING MANUFACTURING TACTICS

Webster defines a *tactic* as "the technique or science of securing the objectives designated by the strategy." The business strategy, derived from a clear understanding of customer needs and the company's position relative to the competition, has driven identification of the strategic emphasis of the manufacturing function. Manufacturing will emphasize some combination of cost, quality, flexibility/responsiveness, or innovation/technology. Achievement of these objectives in turn requires selection of an appropriate set of management actions, or manufacturing tactics.

There are two levels, then, at which a manufacturing function needs to be managed. At a macro, or strategic, level, there are structural issues, such as number of facilities, and their location, focus, and capacity. Degree of vertical integration and choice of process technology and overall organization structure are other structural issues at the strategic level (see Section, "Strategic Decision-Making Categories").

Other issues are managed at a more micro, or tactical, level within the manufacturing organization itself. These tasks can be categorized into three interrelated sets of activities:

1. Control of the manufacturing process.
2. Control of the manufacturing resources including people, materials, and tools.
3. Control of manufacturing information.

Successful manufacturing requires that (1) the right resources (people, materials, tools) be in the right place at the right time, (2) the right processes be executed properly at the right time, and (3) the information to enable and improve execution be available just as needed to enable appropriate, localized decision making. The emphasis in executing any one of these activities, however, must be on the areas of concentration dictated by the manufacturing strategy—cost, quality, flexibility/responsiveness, or innovation/technology.

Manufacturing tactics, then, specify the methods by which process, resource, and information control are executed to optimize their contribution to cost, quality, availability, and features of the end product. This linking of manufacturing tactics, manufacturing strategy, business strategy, and ultimately customer requirements is critical to a company's long-term competitive success.

Over the last decade, a number of concepts have been added to the manufacturing manager's arsenal, including total quality commitment (TQC), just-in-time (JIT), and computer-integrated manufacturing (CIM). Although these methodologies have received considerable attention in both academic and trade media, they do not represent radical change in manufacturing requirements. Rather, they represent modern tools or methods for responding to the old problems of controlling the process, the resources, and the information, respectively. They are evolved from older concepts of quality control, materials management, and line management.

Whether a company chooses to employ TQC, JIT, or CIM philosophies is a choice to be made according to customer need. If any of these programs are implemented, they should be done with a focus on the cost, quality, availability, and features objectives suggested by the business plan.

Let's return to our HP example once again (Figure 3–6). In the instrument business, the R&D/manufacturing link is the most critical to managing the introduction of features required by the customer. In the minicomputer business, TQC programs (as well as the R&D/manufacturing link) are important to meet quality and features needs. In the per-

FIGURE 3–6
Manufacturing Strategic Emphasis

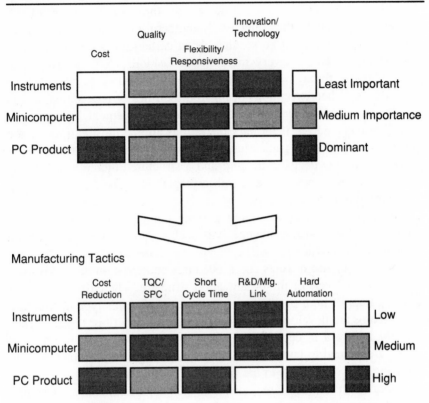

Manufacturing Tactics

sonal computer business, hard automation is required to minimize cycle time at the lowest possible cost, thus allowing customer requirements of low cost and instant availability to be met.

Bottom line? Your choice of manufacturing tactics must be directly linked to customer needs for cost, quality, availability, and features.

ORGANIZE FOR MANUFACTURING SUCCESS

At all levels, the manufacturing organization must be structured and managed to support execution of the business and manufacturing strategies. Volumes have been written[7] on the importance of matching organizational structure with strategy, yet most organizations are poor at explicitly recognizing the linkages.

To begin, the overall structure of the manufacturing organization must match the selected manufacturing strategic emphases. Organizations that must be predictable in certain types of environments can be organized in a very hierarchical fashion. They can run themselves largely through application of clear, well-understood rules and procedures. When exceptions occur to these rules or procedures, a decision request can be passed up through the management hierarchy and direction received. U.S. military operations are the most extreme example of this type of organization. Using strict rules and procedures to guide day-to-day decision making and a structured hierarchy for giving direction during times of stress, military operations operate in reasonably well-structured environments.

Companies concerned about gaining competitive advantage in an environment of constant change will seek a more fluid organizational structure. Such structures will typically be flatter and leaner. They will rely less on the use of rules and procedures and more on decisions made at the lowest levels of the organization structure. The currently popular development of high-performance work teams fits this type of thinking. Work teams are developed at the level of the factory floor to get workers directly involved in decision making about the process they run. Such teams are trained to handle variability and change on their own without referring rule exceptions to the hierarchy for resolution.

Organizations that are more rigidly, or hierarchically, structured will often handle a specific change by creating a suborganization with a more fluid structure. IBM was able to rapidly introduce their new personal computer product by creating a small, independent development organization in Florida. By doing so, they facilitated the communication required to rapidly and successfully launch the new product. The Japanese have used similar spinoff types of structures to introduce new products from cameras to cars. A fitting match of the organizational structure to the current strategic needs has proved crucial in all of these cases.

Once the organizational structure has been put in place, the performance measurement and capital justification systems must be designed both to match the organizational structure and to encourage the behavior desired by the manufacturing strategic objectives.

Performance measurement systems play a particularly important role in the execution of a strategy. Will you measure your managers on achievement of cost objectives (the bottom line) even when time-to-market is your critical success parameter? Many companies do. Current popular literature on developments in cost accounting[8] recommends significant changes in our traditional approaches to cost roll-up, product pricing, and thus performance measurement. It has been said, "Tell me how you'll measure me and I'll tell you how I'll act.". Performance measurement systems *must* be designed with your strategic direction in mind, or it is unlikely that your strategic objectives will be met. The design of such systems is not well understood today. The systems you design will require constant feedback and fine tuning until new and more effective measurement systems become widely accepted.

Similarly, the process used for project justification must match strategic requirements as well.[9] The commonly used return-on-investment approaches are fundamentally cost-focused. They don't directly accommodate inclusion of quality, availability, and feature improvements, either short term or long term.

Quality projects should be measured on contributions to quality; availability projects, on improvements in availability; and production process innovation projects, by their impact on feature enhancement. The application of short-term economic measures to all manufacturing projects is insufficient. New methods of project justification must be found and used by companies wishing to optimize variables other than cost in their business strategy.

The manufacturing organization must be structured to meet the strategic needs of the business, taught the strategic objectives involved in the investment decisions that express the functional strategy, and then measured on its ability to meet those objectives.

MEASURE THE RESULTS

How can a manufacturing business ensure that its strategy becomes more to the organization than just a document that collects dust on the bookshelf? It must provide a system of checks and balances and feedback loops that allow constant monitoring of progress against the strategic objectives.

First, if a strategy is really made up of the patterns of decisions that an organization makes, one ought to be able to check its execution by simply monitoring decision patterns over time. A major food processing corporation we worked with stated that its strategy is to be responsive and flexible in its manufacturing function. Examination of its capital spending plans, however, revealed that over 80 percent of the budget was being approved for cost containment projects. This pattern of decisions suggested that a different strategy—cost reduction—was actually being executed. Watch the decisions that your manufacturing management makes over time and determine whether your strategy is being executed as defined.

We discussed performance measurement at length in the Section, "Organize for Manufacturing Success," as an element of organizational design. It bears repeating here that the performance measurement system must be restructured to support execution of the strategic objectives. To be most effective, the system must also provide competitive norms to reinforce the strategic objectives. The system must reflect contributions to quality, availability, and features, as well as to cost, or the organization will focus on cost to the exclusion of the other factors. Recently, the CEOs of both Hewlett-Packard and Motorola have made quality a primary objective for their organizations and have achieved significant improvements.

In 1979, John Young, HP's CEO, challenged his management teams to improve the quality of the company's hardware products tenfold in the upcoming ten years. This "stretch" objective for the eighties has fundamentally altered HP's management style. HP had always been known for high product quality, but it was experiencing an initial in-process failure rate of roughly 3 percent on a number of products. By 1985, HP engineers had achieved three orders of magnitude improvement in their printed circuit board manufacturing quality, providing first-pass failure rates in the 30 parts per million range. Through 1988, company performance is on track, and the quest continues.

Motorola's Chairman, Robert Galvin, established a similar objective, the result of which was the award of the Malcolm Baldridge Award for Quality to the *entire* Motorola Corporation. In both of these examples, the companies have succeeded in meeting their CEO's objectives only by establishing quality as a performance measure at least equal in importance to any cost objectives that already existed.

In short, strategies can be made to live only in organizations in which the performance measurement systems appropriately reflect primary customer needs, and the competitive response and investment management decisions are made with the strategic direction in mind.

FIGURE 3–7
Five-Step Strategic Planning Process

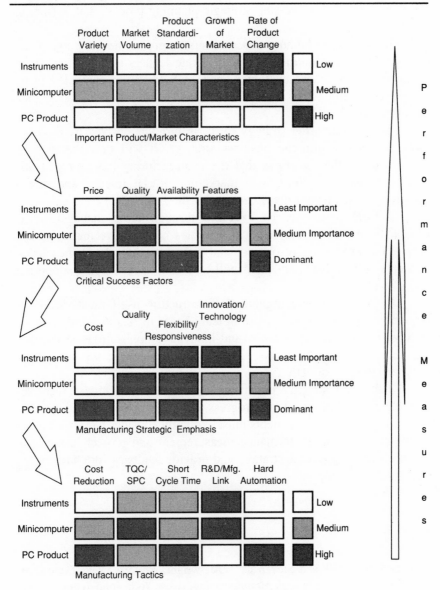

Important Product/Market Characteristics

Critical Success Factors

Manufacturing Strategic Emphasis

Manufacturing Tactics

SUMMARY

We have proposed a five-step strategic planning process that is summarized in Figure 3–7.

The strategic planning process must begin at the very top of the company with commitment from corporate leaders to a focused set of customer needs. These customer needs will likely be contained in the categories of product or service cost, quality, availability, and product or service features.

The manufacturing strategy must, in turn, be derived by the manufacturing team from the business strategy. It will describe in more detailed terms the directions that the manufacturing function will take in playing its role, along with marketing, R&D, finance and human resources, in achieving the objectives of the business.

Strategies can only be executed through close management of tactics. In manufacturing, tactical activities will involve managing and controlling processes, resources, and information. Perhaps the current concepts of TQC, JIT, and CIM will be useful in identifying supportive sets of tactics.

The manufacturing organization's structure itself must match the requirements placed upon it. Use of team-based management techniques, creative compensation systems, and innovative applications of communications technology may be techniques worth considering as required by your strategic direction.

Finally, no strategy can be deemed successful unless there are tangible assessments made of progress against the objectives dictated by the strategy. These assessments must explicitly consider the performance of the competition. Performance measurement and project justification systems must support the strategy and provide feedback for changing the strategy as needed.

NOTES

1. Robert H. Hayes and Steven C. Wheelwright, *Restoring Our Competitive Edge: Competing Through Manufacturing*, (New York: John Wiley & Sons, 1984), pp. 29
2. Ibid.
3. Roger W. Schmenner, "Look Beyond the Obvious in Plant Location," *Harvard Business Review*, January–February 1979, pp. 126–132.

4. Robert H. Hayes and Steven C. Wheelwright, "Link Manufacturing Process and Product Life Cycles," *Harvard Business Review*, January–February 1979, pp. 133–140.

5. Ira C. Magaziner and Mark Patinkin, "Fast Heat: How Korea Won the Microwave War," *Harvard Business Review*, January–February 1989, pp. 83–92.

6. Wickham Skinner, "The Focused Factory," *Harvard Business Review*, May–June 1974, pp. 113–121.

7. See, for example, J. D. Thompson, *Organizations in Action*, (New York: McGraw-Hill Book Company, 1967); P. R. Lawrence and J. W. Lorsch, "Differentiation and Integration in Complex Organizations," *Administrative Science Quarterly*, 12 (June 1967).

8. See, for example, Robin Cooper, "You Need a New Cost System When . . . ," *Harvard Business Review*, January–February 1989, pp. 77–82.

9. Jack R. Meredith and Marianne M. Hill, "Justifying New Manufacturing Systems: A Managerial Approach," *Sloan Management Review*, Summer 1987, pp. 49–61.

CHAPTER 4

STRATEGIC ANALYSIS FOR GLOBAL MANUFACTURING

Linda G. Sprague

Editor's note

This chapter shows how a global manufacturing strategy can be fully developed to complement overall business objectives. Dr. Sprague's process of manufacturing strategy development requires completing a series of planning grids, followed by capacity and technology analysis. As more information about manufacturing capabilities is plugged into the planning grids, the production function's role becomes that of a driver, rather than a follower, of strategic decisions.

MANUFACTURING STRATEGY — THE MISSING PIECE

Most strategic plans contain a barn-door size hole. This alarming fact is disguised by the hefty weight of documents that feature comprehensive analyses of markets, segments and niches, products old and new, competitors' likely market postures, and elaborate financial projections. Occasionally, one finds a brief section toward the end devoted to "implementation," which speaks of "investment in high-technology approaches to meet compelling market challenges and opportunities." When the strategic plan encompasses world-wide markets, the implementation section may refer obliquely to "exploitation of existing capacity close to existing markets."

What's missing is the manufacturing strategy necessary to execute the marketing strategy.

Strategic planning for manufacturing capacity remains the missing link in corporate strategy nearly two decades after the first clear alerts were sounded by Wickham Skinner in 1969. It is serious enough when business strategies for domestic activities are concerned; it can be life threatening when global plans are being developed.

Strategic analysis for long-range business planning typically revolves around markets and products. A widely used structure is based on three dimensions for business strategy formulation:

- Customer groups served.
- Customer functions served.
- Technologies used.[1]

Figure 4–1 is an example of the classic use of this framework for the photo-typesetting industry.[2] The vertical axis identifies customer groups or market segments—in this instance newspapers, commercial printers, typesetting houses, and businesses doing typesetting in-house. The horizontal axis identifies functions or benefits desired by the customer—here speed, quality, and flexibility. Technologies are shown as the third dimension—for the photo-typesetting industry these are generations through which the product technology has evolved and continues to evolve. The photo-typesetting technology generations are:

I. Letter matrix through fixed lens, to film, and to plate.
II. Letter disc through interchangeable lens, to film, to plate.
III. Computer library through CRT, through lens, to film, to plate.
IV. Computer library through CRT to plate.
V. Computer library to laser printer.

This product technology is evolving from mechanical/ photographic, through mechanical/optical/electronic/ photographic, to pure computer-based electronics. The implications for required process technologies in manufacturing are profound. A decade ago, being in this business meant having mechanical production capabilities. Today it means having to work successfully with computer and laser technologies: The older technologies are diminishing in importance.

FIGURE 4–1
Three-Dimensional Structure for Business Strategy Development

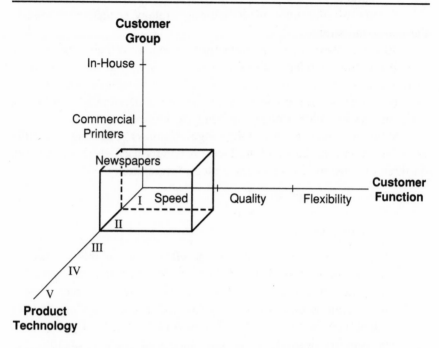

Strategic planning using Abel's three-dimensional structure can lead to understanding the approaches to market segments with subsequent development of specific market strategies. As the product technology develops and evolves, market strategies can be refocused, and in some instances product technology developments can be anticipated.

For example, the small cube in Figure 4–1 describes the approach that was successful in selling second-generation photo-typesetting equipment to the newspaper industry on the basis of speed. Third generation equipment added the potential for higher quality and greater flexibility, making it attractive to commercial printers; this segment can also be represented in Figure 4–1. Thoughtful use of this framework can play an important role in the development of business strategies in industries facing substantial product technological development.

However, from the perspective of the manufacturing and logistics arms of the organization, a view of business planning based on only three dimensions (customer groups, function, and technologies) is woefully incomplete. The process technologies required to create the product technologies are left in the shadows. As the example in Figure 4–1 shows, product technology development can, and too often does, force dramatic change in process technology requirements.

The evolution of photo-typesetting product technologies has meant a shift from the manufacture of electromechanical devices toward the production of software-driven, computer-based systems. The impact on manufacturing capacity planning—for both physical technology and work force expertise—has been wrenching. The frameworks used for analysis of such change are incomplete; they do not go the next step and add the dimension of process technology. And, when global marketing and distribution is involved, this framework must be augmented by geography.

Without a fully conceived and developed strategic program supporting manufacturing and logistics, effective implementation of a business strategy is impossible. The track record for implementation of corporate strategy has been spotty. One reason may be the fact that most strategic planning is truncated. The hole in the plan is the lack of manufacturing and logistics strategy development as part of the corporate strategy.

HOW TO DEVELOP STRATEGIC PLANNING GRIDS

The first step in plugging the major hole in strategic planning is to integrate manufacturing and logistics into the process. We can begin by laying out a simple grid. Then we will add dimensions to the basic grid, building a set of grids that fully describes the problem.

Let's start with Figure 4–2, in which one side represents the marketplace, and another the product. We use the example of three types of lawn mowers (nonpowered, powered, and riding mowers) used to teach the fundamental marketing concepts. The marketplace is broken down into three segments—city, suburban, and rural. Forecast volumes are entered into the cells in this grid. This grid is the basic framework commonly used by marketers to identify and explore specific product opportunities or market niches.[3]

Describing the three types of lawn mowers does not fully tell the technology story: The powered mowers are based on two types of power, gasoline or electricity. Figure 4–3 adds the product technology dimension

FIGURE 4-2
Market/Product Grid

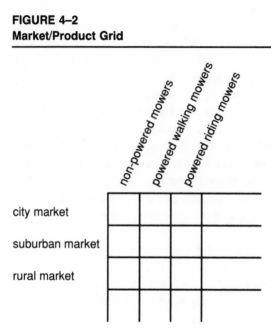

that will move us closer to the critical issues facing manufacturing—the process technology requirements and options.

Global marketers will expand the market axis to show geographic location. The addition of this international dimension is shown in Figure 4-4. Finally, in Figure 4-5, the key manufacturing axis is shown—process technology requirements (injection molding, painting, etc.). Bringing this final dimension into focus provides the basic information necessary for the development of a complete strategic plan.

Step 1: The Market/Product Grid (Figure 4-2)

The market/product grid sets the foundation for the development of the marketing strategy. For an international organization, the market segments are further specified by country. Present and projected sales volumes are typically shown in each cell in order to provide some sense of time, essentially adding a third dimension to the basic grid.

The information provided by such market analyses is of fundamental value for the business-strategy planning process. This basic framework can be considerably enriched with segment and niche details and by delineation of complete product lines. However, completion of Step 1, The Market/Product Grid, is not sufficient without the next part of the process—the development of complementary manufacturing and physical distribution segments of the strategy. Without these vital next steps, the strategic plan, for all its bulk, will be incomplete.

Manufacturing and logistics organizations within the firm need information provided by a systematic conversion of base-line market forecast data into projections of specific product technologies and their relative importance over time. From this information, manufacturing sees the resulting implications for required production technologies. From a market forecast for engines we can develop capacity requirements for forging crankshafts or assembling carburetors. To complete the strategic analysis for an international organization, the geographic locations of capacity availability, present and projected, must be added.

Step 2: The Market/Product Technology Grid (Figure 4–3)

The product delineation shown in a traditional market/product grid is based on customer-focused product differentiation: The objective is market niche analysis. Differentiation of products by their technology may result, but it is not the point of the exercise. The lawnmower analysis grid is a good example of this. As it happens, these three products (nonpowered, powered walking, and powered riding) differ substantially with respect to their technologies, although market segmentation was the basis for the distinction.

The first step in the development of the market/product grid into a tool for operations analysis and planning is to add product technology factors to it. To continue with the lawn mower analysis, the grid must be broken down further into gas or electric powered engines.

Not surprisingly, rural markets show the biggest demand for riding mowers. If that demand should shift to suburbia, capacity requirements, location of demand, and logistics (warehousing and transportation costs) would also change. This becomes clearer when each cell contains both present and projected volumes. There are also clear implications for research and development priorities for products. Anticipated further population movement back into cities, for example, would probably dictate lower gas riding mower forecasts, although there would also be a need for new products in the urban market. This information can

FIGURE 4–3
Market/Product Technology Grid

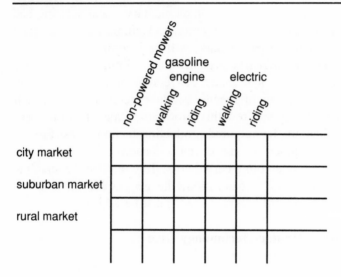

also be of use for additional market analyses, particularly if markets are segmented geographically.

Working through the completion of the grid in Figure 4–4, with its geographic and technology volume forecasts, may allow us to spot a profitable opportunity, for example, in the demand for older product technologies in lesser developed parts of the world. This framework provides a mechanism for identifying such opportunities.

The projected volumes by product technology also have implications for future after-sale service and accessory requirements. This business has received new attention from U.S. suppliers because its profit margins are usually multiples of the original equipment margins. The extra revenue does not come easily, however. The service business is difficult to translate into capacity requirements. The product line needs to be taken apart first, separating purchased from in-house parts.

Step 3: The Product Technology/Process Technology
Grid with a Strategic Capacity Requirements Explosion

U.S. manufacturers need to understand shifting capacity requirements. With export demand pumped up by the low value of the dollar, it is

FIGURE 4–4
Geographic Market/Product Technology Grid

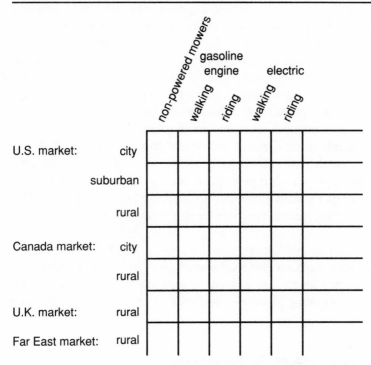

dangerous not to understand the capacity issue thoroughly in all its forms. Specifically, analysis of true capacity requirements allows us to consider various strategic responses:

1. We can consider gearing up or down quickly and define relevant costs/paybacks in advance.
2. We can compare the costs of increasing different capacities: Are we just adding a shift of hourly workers, for example, or would we be looking at renting and equipping a new facility?
3. We can compare the effect on lead times of each strategic response in capacity planning.

Long-range changes in capacity needs implied by shifts in products and their technologies can be identified on the product technology/process technology grid. This is the critical link between market strategy development and strategic operations planning. For lawn mowers, the grid looks like that of Figure 4-5.

It is common for manufacturing organizations using their material requirements planning systems to carry out requirements explosions for parts and for elements of capacity. This is typically accomplished as part of a relatively short-range program of execution, not as part of a strategic planning process. The short-range focus, although highly valuable for day-to-day operations and control, cannot provide visibility for long-range needs for new production technologies.

For example, material requirements planning/driven capacity planning may generate good demand data short-term on total circuit board requirements, but it will not tell us if and when a technologically very different board type will shift process/capacity plans (e.g., moving from single to multilayer boards). Nor can it illuminate the potential for older technologies whose future competitive importance may be underestimated.

The product technology/process technology grid is essentially a strategic capacity requirements explosion with each cell containing present and projected levels. It is based on information too often ignored in the strategic planning process—the forms of process technology cur-

FIGURE 4–5
Product Technology/Process Technology Grid

	machine shop	injection mode	sheet metal	fiberglass	welding	painting	electrical	circuit boards	purchasing	
non-powered mowers										
gas engine: walking										
riding										
gas engine: walking										
riding										

rently in use and estimates of those required in the future. The estimates are not easy to develop, in part because alternatives are often possible, particularly with respect to the purchase or subcontract of some processes.

Unfortunately the information within these cells, both present and projected, is not easy to gather. At the least, a bill of material explosion capability is required, along with specific process data (routings) and utilization data (run times by machine or work center, setup and change-over data). Labor utilization data can also be factored in. Where a fully developed capacity-requirements planning system is available, organization of base data for the present situation is fairly straightforward. A firm that does not have this MRP (material requirements planning) capability will find conversion of product demand into meaningful capacity load numbers to be a daunting, although worthwhile, task. Even when the information about present process capacities and utilizations is at hand, developing estimates of future technology requirements is bound to be difficult. We usually see an overabundance of short-range data coupled with lack of longer-range information about technological processes.

Alternative technologies, or approaches to technologies, pose a particular problem in forecasting process requirements. Purchased parts provide a good example of the kind of issues that become apparent during the development of process capacity requirements projections. For example, if a firm purchases all of its circuit boards today because analysis has shown that insufficient volume exists to produce them economically in-house, simply extending current values of purchased parts into the future will mask opportunities for economic production as volumes increase.

Incorporating forecasts of future process economics is important as well. To continue the circuit board example, as breakeven volumes for economic production of boards drop, the parameters for make/buy decisions are changed.

The product technology/process technology grid makes clear aggregate volume needs by process, yielding baseline data for consolidation. When each cell in the grid contains present and projected volumes, the information can be reorganized to show changes in overall capacity requirements as well as shifts in process mix requirements.

The capacity requirements projection (Figure 4–6), for example, shows that in ten years the product will require work in welding, but

FIGURE 4–6
Present and Projected Process Technology Requirements

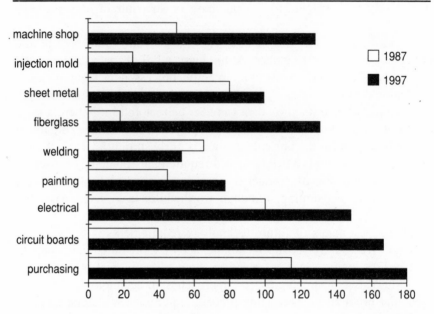

significantly more (proportionally) in new technologies, fiberglass, and circuit boards.

Analysis of this information results in estimates of aggregate capacity needs (the size of the building) and the relative needs of process technology (the relative sizes of the process centers within the building). It can also highlight research and development needs for process technologies as well as long-range requirements for technical expertise.

To bring these grid analyses full circle, process technologies can be mapped against geographic location, present and proposed. When studied along with a geographic market/product technology grid (Figure 4–4), the maps will highlight manufacturing location options that offer the best support for the competitive situation. The geographic grids also provide baseline information for the development of long-range logistics and after-sale support programs to complement the marketing strategy.

Strategic planning for global markets and the manufacturing and logistics to support effective presence in those markets require planning information along these dimensions:

- Market by segment.
- Product by product line.
- Geographic market.
- Product technology.
- Process technology.
- Geographic process availability.
- Time.

We have used grids that are basically two-dimensional plots of multidimensional information. Rotations of the axes bring information pairs into focus, permitting systematic analysis of the implications of product technology shifts on process technology requirements. Just as the market/product grid identified trade-offs in the strategic marketing process, the product technology/process technology grid (Figure 4–5) facilitates trade-offs in strategic manufacturing and logistics planning. The addition of geographic location information makes this a particularly valuable tool for the planning of global manufacturing capability.

STRATEGY: PRODUCT-DRIVEN OR PROCESS-DRIVEN?

Most organizations assume that product technology requirements will drive manufacturing process requirements. This is often the case, but process technology advantages can be exploited for competitive market position, particularly when a low-cost market opportunity opens up. Although the sequence we have used is the one textbooks traditionally show, with the market-product grid as the pacesetter, there are situations when the driving force can be the process technology advantage. Process technologies that provide the ability to deliver a customized product in a matter of days, for instance, can offer a major competitive advantage.

The personal computer clone business is an example of competitive advantage through fast delivery of customized units. A mail ordered system, for example, can be delivered from Wisconsin in five working days (the actual final assembly takes about half a day). The basic system is made up of one Seagate 30 megabyte hard disk, one 12.5 megahertz 80286 chip CPU motherboard, one Adaptec controller, one TEAC floppy disk drive, one power supply, and one case—all pieces currently available on the market. The product competes with other IBM-

AT clones that are twice as expensive. It fills a niche created by a rare IBM marketing strategy error. When IBM decided to move the market away from their five-year old AT design, they failed to note the numbers of users who were satisfied with the IBM design, but not the price, and who were not eager to move up to their next product offering, the System 2.

WHERE TO GET THE DATA

The real problem in the development of these grids is data availability. Present market and product line detail is the most likely data to be available. Projections—particularly of new product and product technologies—will of course be forecasts with the usual accuracy problems.

The data conversion process is deceptively simple. It requires a file inversion. Data is collected by product, and exploded through the material requirements planning system to component parts. The data must be manipulated into meaningful capacity categories. Conversion of demand to load tends to be the least well done, so the result is that we often load facilities by dollars. It doesn't work!

The first grid shown (Figure 4–2), the market/product grid, can generally be drawn quickly with readily available data from current reports and regular marketing planning documents. Beyond that, data availability in the form required for these frameworks becomes a serious problem, particularly when international sales and production are involved.

An added complexity is the difficulty of multicurrency conversion. We can expect that world-wide currencies will continue to fluctuate. Strategic planning in the face of these paper fluctuations will be difficult.

Further, currency complications are compounded by the peculiarities of cost accounting systems, those used in the United States included. The important caveat for this or any other exercise intended to shed light on aspects of global manufacturing and logistics is to avoid making long-term decisions based on short-term oddities.

Data-availability problems notwithstanding, these grids, which focus attention on process technology requirements, provide a window on future needs, conventional and innovative. Complementing market-based strategic planning frameworks with others based on manufacturing and logistics requirements results in business strategy planning with a higher probability of successful implementation. It can also highlight competitive options focused on manufacturing and logistics strengths. The barn-door size hole can be closed.

NOTES

1. D. A. Abell, *Defining the Business: The Starting Point of Strategic Planning* (Englewood Cliffs, NJ: Prentice-Hall, 1980).
2. From data in "Bobstgraphic Division (A)," case prepared by Nancy Newcomer, Research Associate, under the supervision of Professor John A. Murray, IMEDE, Lausanne, Switzerland, 1980.
3. E. N. Berkowitz, R. A. Kerin, *Marketing* (St. Louis, MO: Times Mirror/ College Publishing Company, 1986).

BIBLIOGRAPHY

Abell, D. A. *Defining the Business: The Starting Point of Strategic Planning*, Englewood Cliffs, NJ: Prentice-Hall, 1980.

Berkowitz, E. N. , R. A. Kerin, and W. Rudeluis. *Marketing*. St. Louis, MO: Times Mirror/College Publishing Company, 1986.

Heckman, C. R. "Don't Blame Currency Values for Strategic Errors." *Midland Corporate Finance Journal*, New York: Stern Stewart Management Services, Inc., Fall 1986, pp. 45–55.

Porter, M. E. "From Competitive Advantage to Corporate Strategy." *Harvard Business Review*, May–June, 1987, pp. 43–59.

Skinner, W. "Manufacturing—Missing Link in Corporate Strategy. " *Harvard Business Review*, May–June 1969, pp. 136–145.

CHAPTER 5

CHARACTERIZING YOUR ENVIRONMENT—STRATEGIC PLANNING SYSTEMS: IS MORE NECESSARILY BETTER?

Jeanne Liedtka

Editor's note

Most U.S. manufacturing companies operate in a climate of uncertainty and change. This chapter outlines a strategic planning approach that categorizes industry and company factors in four groups of varying degrees of complexity and uncertainty—the Dinosaur Bowl, Mission Control at Miami, Nintendo Warriors, and Dogfight-Aerial Combat—and suggests appropriate planning methods for each.

What are the implications for manufacturing managers trying to match a firm's strategic planning process with its environment? A key issue is that, after all the elaborate analyses have been completed, it is very often manufacturing that determines whether a strategy will succeed. A manufacturing organization can be structured by appropriate use of technology and simple, efficient planning methods to deal effectively with environmental pressures such as uncertainty and complexity. Methods that facilitate dealing with change (shorter cycle times, setups, order processing, and flexibility) are more fully explored in Joe Blackburn's piece on time-based competition and Sara Beckman's chapter on flexibility.

In the twenty years since the concept of strategic planning appeared in the business literature and subsequently took hold in the minds of corporate America, the assumption has been that if *some* planning was good, *more* planning must be even better. That assumption has begun to be seriously questioned of late, with the growth of interest in, for example, entrepreneurship and intrapreneurship. What price do we pay for our carefully integrated strategic plans with their long time horizons? More than just the cost of the bureaucracy of staff planners who maintain them? Is what is good for General Motors generally good for the rest of American business?

For too long manufacturing managers have been willing to relegate involvement in strategic planning issues to senior management and staff planners while they focused exclusively on issues such as product quality and production efficiency. This is a luxury they can no longer afford in an atmosphere of escalating uncertainty resulting from globalized competition, technological innovation, and demographic changes in the work force. With the least flexibility to change and the longest lead times for capacity additions, manufacturing would benefit more than any other company division from an accurate crystal ball. Increasingly, the production efficiency of tomorrow rests on today's ability to put in place the proper internal capabilities to respond to new developments, whether they be in customer demand, product or manufacturing technology, or competitors' behavior.

Yet, business firms operate in environments that differ quite dramatically from each other. Thus, "cookie cutter" techniques are unlikely to produce satisfactory solutions across the board. The first step toward putting in place a successful process for making important strategic decisions for your firm is to take a close and careful look at your environment. This has important implications for the *type* of planning process you use, as well as for the ultimate competitive strategy that you select.

CHARACTERIZING A FIRM'S ENVIRONMENT

There are many ways to describe the different types of environments that firms operate in. None can be considered "right" or "wrong" since each focuses on a particular aspect or characteristic. Some, however, have proven to be more useful than others.

Michael E. Porter, a professor at the Harvard Business School, has developed a very popular framework that focuses on the characteristics of the industry that a firm operates in.[1] An industrial economist by training, Porter argues that the range of profitability within an industry is determined by five structural characteristics.

The first is the *threat of new entrants*. If an industry is easy for new firms to enter, Porter argues, profitability will be decreased in the price wars initiated as they attempt to gain market share from already entrenched competitors. Deregulation, for instance, facilitated the entry of numerous new carriers, such as People Express, into the airline industry. In the price wars that followed, profits plummeted.

The level of *rivalry among existing firms* is another important factor in Porter's scheme. Intense rivalry is likely to destabilize the industry, with much the same consequences as the entry of new competitors. Structural factors such as high fixed costs, lack of product differentiation, and slow growth all contribute to increased rivalry as firms scramble to fill excess capacity through the use of price cuts. The paper industry in its maturity offers a case in point with its bitter rivalry and low profitability.

The *threat of substitute products*, a third factor, must be assessed by looking outside a firm's own industry at products that perform a function similar to one's own. The ready availability of alternatives for the customer tends to reduce profits by placing an upper limit on the price that can be charged, even in times of shortage. Firms often make the mistake of defining their competition too narrowly, of ignoring new developments outside of their industry, which may later lead to erosion in the demand for their own products. The use of personal computers to perform word processing, for instance, has annihilated demand for traditional typewriters in a way unforeseen by most of the industry even ten years ago.

Finally, *the bargaining power of suppliers and buyers*, respectively, constitute the final two important structural characteristics of industry in Porter's theory. At one extreme, powerful suppliers are likely to demand top dollar for raw materials, while at the other extreme, powerful buyers resist a firm's attempt to pass these cost increases along to them. The result: decreased profitability for the industry caught in between. Strength often comes from concentration. Consider the hypothetical case of a small manufacturer of automobile windshields in Detroit. Forced to accept substantial price increases from a small group of glass manufacturers who control the supply of raw materials, but unable to pass these increases along to the big three car companies (each of whom constitutes too substantial a portion of his business to lose), the windshield manufacturer suffers the loss of profits.

These five factors, Porter argues, have important implications for the selection of one of the generic strategies through which firms compete with each other. These strategic alternatives include:

- Cost leadership (being the competitor with the lowest cost).
- Focus (targeting a particular segment of the market to serve).
- Differentiation (distinguishing oneself from one's competition via some unique capability, e.g., superior quality, customer service).

or some combination of these three.

Though Porter's view of the environment is intuitively appealing and very useful in helping managers think about the *content* of an appropriate strategy, it does not help us address *how* to plan or illuminate the internal *process* of arriving at a strategy within the firm.

COMPLEXITY AND UNCERTAINTY, TWO ADDITIONAL PLANNING FACTORS

Researchers in the field of organizational theory have generally used broad concepts in describing the qualities of the environment to which an organization ought to respond in designing its internal processes. The two aspects usually focused on have been complexity and uncertainty, the most critical determinants of a firm's planning process. Appendix I discusses some common measures that have been used as indicators of the degree of environmental complexity and uncertainty.

Complexity

Complexity refers to the number of different factors in a firm's environment to which it must pay attention. These things are often related or interconnected in some way. The term "complexity" is frequently used interchangeably with other terms such as "heterogeneity" or "diversity."

The complexity of a firm's environment is a function of both internal and external factors. For example, a firm with numerous product lines probably operates in a more complex environment than a single-product firm. It has more customers' needs to be attentive to, more competitors' actions to monitor, more inputs to analyze and integrate. A change in any one factor (customer needs, for example) is likely to have a ripple effect on other factors (e.g., competitors' behavior). Thus, a firm that operates in a more complex environment needs to spend proportionately more time monitoring, or "scanning" its environment for new developments. Firms that operate in more homogeneous, less complicated environments can afford to do less of this. Consider, for example, the situation of a large, multiproduct firm like General Electric, which has made products as diverse as lightbulbs and nuclear generating equipment. Clearly the complexity of the firm's operations requires that it devote substantial effort, in terms of manpower, to tracking new developments in all of its many areas of participation.

Uncertainty

Uncertainty, our second important quality, relates to the degree of change in a firm's environment, and to the frequency, extent, and (most importantly) predictability of that change. The more frequent, significant, and unpredictable a firm's environment is, the more uncertain it is. The term "uncertainty" can be used interchangeably with turbulence, dynamism, and instability. Operating in a field like consumer electronics, characterized by frequent and significant technological change, a firm such as Matsushita would be considered to operate in an environment high in uncertainty. Proctor and Gamble, on the other hand, a firm with a more stable technology in the household products area, would experience more certainty.

Though these descriptions of the environment may seem self-evident at first, they have important (and often unnoticed) implications for how firms make important strategic decisions. Rating each of the two qualities low or high for a given company (see Figure 5–1), we would find that firms operate in one of four possible environments.[2]

After a discussion of planning processes in general, we will return to each of the four quadrants to explore the impact of these very different environmental contexts.

FIGURE 5–1
Simplified Matrix for Identifying a Firm's Strategic Planning Environment

	Low complexity	High complexity
Low uncertainty	I	II
High uncertainty	III	IV

DIMENSIONS OF THE PLANNING PROCESS

Having looked at the characteristics that we will use to describe the firm's environment, we now turn to a discussion of the dimensions of the planning process itself. Four different aspects of the planning process appear to be related, ultimately, to the context in which the firm finds itself.

Degree of Centralization

The first of these is the degree of centralization of the planning process— whether it is primarily "top-down" versus "bottom-up. " This refers to the level in the organization where the alternatives for strategic action are generated and evaluated. In a top-down process, senior management is primarily responsible for strategy formulation. Their decisions are then passed down through the ranks to middle- and first-line levels of management for implementation. Conversely, in a bottom–up process, lower levels of management have an important voice in creating and selecting alternative strategies, as well as for implementing them.

Who Makes the Strategy, Line or Staff?

Another dimension of the planning process, sometimes related to the degree of centralization, is the role of staff in relation to line management in the formulation of strategy. In some organizations, strategy is formulated largely by a staff of planning experts, usually located at corporate headquarters. These may be employees of the firm or outside consultants hired from major consulting firms such as McKinsey, the Boston Consulting Group, or Bain and Company. In other firms, strategy formulation is the responsibility of line management—of the managers who actually run the business on a day-to-day basis.

The first two dimensions, then, relate to *who* in the organization generates the firm's strategy. The final two dimensions deal with the nature of the process itself.

Comprehensiveness of Process

The third dimension examines the comprehensiveness of the process. The comprehensiveness dimension is concerned with how exhaustive and extensive the process attempts to be. It encompasses such specific factors as the time horizon of the process, as well as the linkage of the planning process to other organizational processes such as budgeting, performance appraisal, and management incentives. Increasing comprehensiveness also results in increased formality in the process, implying additional paperwork, deadlines, and levels of review. These are usually necessary in a decentralized organization in order to efficiently process and integrate numerous inputs. To illustrate the differences in the degree of comprehensiveness, we might contrast the very different processes involved for the entrepreneur who decides, "Let's go for it," based on gut feelings, versus the large, multinational firm whose strategic plan, and the analysis supporting it, result in a 200-page document reflecting years of effort.

Impetus Driving the Process

The final aspect of the planning process that we will concern ourselves with is the firm's analytic orientation. By *analytic orientation*, we mean the focus of and the impetus behind the process. At the extremes, a firm's process can be either internally focused and reactive in nature or externally focused and proactive in nature. The focus of an internal, reactive orientation is usually to *solve problems* that have arisen and to emphasize the gathering, analyzing, and generating of data on the firm itself (such as revenue and earnings-per-share targets) as the basis for strategic decision making. As a result, the firm usually adapts *after the fact* to changes in its environment. In contrast, the external proactive orientation is focused primarily on the search for new opportunities. In this process, the gathering of outside data dominates (market growth rates, assessments of competitors' behaviors, etc.). The goal here is to manipulate the firm's environment, or strategy, in *anticipation* of a change occurring.

Linking a Firm's Strategic Planning Process to its Environment

Having defined, thus far, two important characteristics of a firm's environment (complexity and uncertainty), and four dimensions of the strategic planning process (degree of centralization, role of staff versus line management, degree of comprehensiveness, and analytic orientation) we arrive at the critical question: Do these relate to each other in a way that is significant? The answer is yes, and we begin to understand why as we examine the very different impacts that complexity and uncertainty have as we try to plan long term.

Coping with Complexity

Essentially, complexity is *knowable*; uncertainty is not. In other words, if complexity refers to having a large number of different factors in the environment, it ought to be analyzable and therefore *controllable* if we as a firm devote sufficient resources to paying attention to it. That implies having a very comprehensive bottom-up planning process that has a lot of staff support and an external proactive orientation.

Complex environments require decentralization for good reasons. The more specialized markets a firm participates in, the less knowledge senior management has about any one of those markets. As a result, the detailed knowledge necessary to make important strategic decisions occurs only at lower levels in the organization. Thus, only the managers in the relevant area of specialization are equipped to formulate strategies for their products. Imagine Jack Welch attempting to become an expert on all relevant aspects of each of General Electric's products! That would be inefficient, unworkable, maybe even dangerous to the long-term health of the firm.

Simultaneously, however, all that decentralization requires is a whole staff of people somewhere at corporate level and a comprehensive system with lots of paperwork and deadlines to integrate all of those divisional plans into one coherent plan for the organization as a whole. The process must be formal as a means of control. The diversity of products and markets creates the need for a proactive external orientation. It would be impossible to accomplish corporate objectives in a reactive mode. The potential for inconsistent and incompatible responses is too high.

Coping with Uncertainty

Uncertainty, on the other hand, is inherently unknowable. If a change is rather specifically predictable (the change in the number of 25-year olds between 1980 and 1990, for example), it does not involve much uncertainty. After all, the number of individuals who will be 25 in either 1980 or 1990 have already been born. Consider the greater difficulty in predicting the number of births in 1990. This is substantially less predictable and hence, more uncertain. Regardless of the resources devoted to its analysis, it is difficult to plan for uncertainty. As a result, in an uncertain environment, extensive planning processes may accomplish little.

Unlike complexity, uncertainty fosters centralization, elimination of staff planning specialists, decreased comprehensiveness, and an internal reactive orientation. Centralization offers decreased response time and improved control. Senior management seeks the assistance of technical experts, rather than planning specialists, to selectively monitor certain aspects of the environment (such as technology). The rate of change leaves the firm with little time, substance, or ability to integrate. In this situation, comprehensive or centralized planning systems may be worse than useless; they may have an adverse effect on performance by reducing the organizational flexibility needed to move quickly. Plans with long-time horizons become outdated overnight. Information is obsolete before it reaches headquarters. Formality decreases response time. Worst of all, people become committed to strategies based on assumptions and predictions that turn out to be wrong.

A number of researchers have found a negative relationship between comprehensive planning systems and firm performance in an unstable environment.

Returning to our four-quadrant model discussed earlier, we can now describe the ideal types of planning systems for each quadrant (see Figure 5–2). Remember that analysis of your industry and company environment is the first step. Although life-cycle stages may change, this will help identify where your company fits on the spectrum of relative uncertainty and the spectrum describing relative complexity. Once the company environment has been identified, the amount and type of strategic planning required can be selected. The objective here is to build flexibility and quick response or to generate more detailed information through a centralized structure. Too much planning in a high-tech environment, for example, will probably defeat the whole purpose of the exercise. Too little planning detail in a more formal situation will undermine the credibility of the plan.

FIGURE 5–2
Simplified Strategic Planning Matrix with Paradigms

	Low complexity	High complexity
Low uncertainty	I: The Dinosaur Bowl	II: Mission Control at Miami
High uncertainty	III: Nintendo Warriors	IV: Dogfight - Aerial Combat

Quadrant I: the Dinosaur Bowl

Complexity: Low

Uncertainty: Low

Planning System: Minimal—centralized, reactive; lacking staff planners and comprehensiveness.

Termed the "dinosaurs," few firms remain in this type of halcyon environment. The sweeping changes in both government regulation and technology have rendered extinct the previously stable, monopolistic environment in which they ruled. Once upon a time, lack of complexity here offered little rationale for elaborate planning. The few firms that remain are in previously static environments, (utilities and insurance bureaucracies) whose goal usually is to maintain the status quo. They are interested primarily in problem solving; planning is geared toward internal control, with a short-time horizon. If we looked to the sports world for an analogy, we would find our dinosaurs bowling—same pins, same alley, your own ball every time. Of course, it takes skill and practice, but once you've got it down, the dynamics of the game rarely change.

Quadrant II: Mission Control in Miami

Complexity: High

Uncertainty: Low

Planning System: Proactive, extensively decentralized, comprehensive, including staff planners.

Here, we find the consummate planners—the General Electrics of the world. Generally large, multiproduct firms, the diversity of their operations requires that they decentralize their planning system to facilitate the extensive environmental scanning necessary to formulate an intelligent and coherent overall strategic thrust. Valuing rationality as a dominant goal, the task of coordinating the far-flung planning network requires staff planners who enforce formal standards of reporting to reduce distortions as information flows upward.

We've moved to the Super Bowl, and everyone on the sidelines is wearing headsets to keep up with the action. Life is complex, competition is head to head, and focusing exclusively on your own game is no longer sufficient when you share the field with your competitor. *So you get to know his game as well as your own.* The game is played with carefully planned plays; otherwise you'd have chaos. If you do your homework, however, the strategies of your competitor should ultimately be fairly predictable. The gridiron, unlike the bowling alley, is no place for a team without a carefully conceived, thoroughly researched plan of attack that anticipates the opponents' moves.

Quadrant III: Nintendo Warriors

Complexity: Low

Uncertainty: High

Planning System: As in Quadrant I, decentralized and lacking comprehensiveness, but with technical experts replacing staff planner and a mixed analytic orientation.

Frequently seen in the start-up phases of a new industry like Biotech, where the behavior of both government regulators and potential customers is unpredictable and where the technology itself has not yet stabilized, the combination of low complexity and high uncertainty fosters centralized decision making and little comprehensiveness to ensure a rapid response to environmental changes and maximum flexibility. The action is fast-paced and new developments are unpredictable; the firm must be free to act quickly, unencumbered by layers of bureaucracy and red tape. In order to achieve such minimal response time, decision-making responsibility must rest in the hands of those at the top of the organization. Planning here serves a different focus: It provides a forum through which senior management communicates to employees to calm their fears in the face of vulnerability from external threats and to reinforce the firm's centrally determined mission. As a result, it takes an internal reactive orientation. At the same time, however, this is balanced by the existence of those selected technical gurus, who pay a great deal of attention to a few critical external factors.

This is the type of environment in which the entrepreneurially managed firm thrives. Lean, quick on its feet, the image is of the entrepreneur as Nintendo warrior, with joystick in hand, dodging alien spacecraft on a path guided more by his or her own vision of the future than by the latest statistics on share of market.

Quadrant IV: Dogfight—Aerial Combat

Complexity: High

Uncertainty: High

Planning System: Hybrid—decentralized; low comprehensiveness; external proactive orientation; and the use of technical experts.

The hybrid planning process in this quadrant is forged out of the need to meet the incompatible demands of highly complex, uncertain environments. The high uncertainty suggests the need to allow for rapid response to change while maintaining control. The entrepreneur, joystick in hand, is the analogy we've used.

High complexity, however, forces decentralization. The number of games being played simultaneously is too much for even the most adroit warrior! Time pressures prevent the integration efforts required to synthesize the information derived during the multiple searches for new opportunities being conducted throughout the firm. As a result, control over policy making at senior management levels is effectively forfeited. Industries such as consumer electronics with numerous competitors both small and large, constant innovations in product and production technologies, and changing consumer tastes offer a case in point.

Imagine a basketball game in which each team has their own ball; think of the coach on the sidelines attempting to call plays—or the team members attempting to execute them.

Visualize a dogfight (aerial combat) in progress. The squadron commander may monitor who is winning or losing each pilot's individual battle, but he is unable to tell them *how* to fight. In the heat of the moment, that responsibility must fall to the pilots alone.

Structuring Your Strategic Planning System— Be Selective

What are the implications of these observations for management practice? They are straightforward, in a general sense: *Good strategic planning requires that you begin with an understanding of the environment in which your firm operates and of the role of manufacturing in that strategy.* The types of elaborate and comprehensive planning systems that have routinely come to be associated with good management are undoubtedly appropriate and commendable for some firms, but could potentially spell disaster for others. Like most aspects of the art of managing, strategic planning has few universal rules. Instead, it has a series of prescriptions, with the usefulness of any one of these being contingent on a variety of factors. The ones we've examined here—the levels of complexity and uncertainty in the firm's environment—are certainly among the most important for managers to consider in structuring the strategic planning process most suited to the special qualities of their own environment. The Appendix to this chapter discusses some measures that have been commonly used as indicators of the degree of environmental complexity and uncertainty.

NOTES

1. Porter lays out his theory in his seminal book called *Competitive Strategy, Techniques for Analyzing Industries and Competitors* (New York: The Free Press, 1985). *Competitive Advantage, Creating and Sustaining Superior Performance*, (New York: The Free Press, 1980) follows up at a more detailed level.
2. Obviously, these qualities operate on a continuum from low to high. We have reduced the complexity of the real world into the simplicity of four quadrants to illustrate our point more clearly.

APPENDIX TO CHAPTER 5: DIAGNOSING YOUR ENVIRONMENT

Measures of Complexity

Environments that are less complex tend to be characterized by:	*Environments that are more complex tend to be:*
Single product firms with few subsidiaries	Diversified firms with many subsidiaries
Regionalized competition	Global competition
Little vertical integration	Significant vertical integration
Few competitors	Numerous competitors
Single-channel distribution system	Multichannel distribution systems
Short lead times on component parts	Long lead times on component parts
Few internally shared facilities	Multiple internally shared facilities
Few manufacturing technologies involved	Multiple manufacturing technologies involved
Single-customer base with standardized needs	Differentiated customers with specialized needs

Measures of Uncertainty

Environments that are less uncertain tend to be characterized by:	*Environments that are more uncertain have:*
Stable manufacturing or product technology	Changing manufacturing or product technology
Small and stable group of competitors	Numerous competitors with new industry entrants
Modest product modification activity	Frequent innovation and new product introductions
Good information concerning competitors' strategies and behaviors	Poor information concerning competitors' strategies
Predictable customer demand patterns at the aggregate industry level	Changing customer demand patterns at the aggregate industry level
Secure raw material availability	Questionable raw material availability
Low likelihood of product obsolescence	High likelihood of product obsolescence
Consistent product quality	Erratic product quality
Entrenched, long-standing government regulation	Changing degrees of government interference

PART 3

NEW STRATEGIES

Part Three, "New Strategies," covers all business functions that must be included in the development of manufacturing strategy from the beginning of the order cycle in white collar areas to manufacturing quality control and logistics. The issue of manufacturing flexibility as the key to building more competitive facilities is fully developed in Sara Beckman's chapter. Hal Mather defines other elements of the product design and delivery cycle that must be addressed to choose appropriate manufacturing techniques, including control of product mix. Bob McInturff looks at three success stories from the perspective of Human Resource Management. And finally, the Digital Equipment Corporation chapter looks at how a successful company manages change.

CHAPTER 6

MANUFACTURING FLEXIBILITY: THE NEXT SOURCE OF COMPETITIVE ADVANTAGE

Sara L. Beckman

Editor's note

What is flexibility? Why should United States manufacturers become more competitive in this area? In this chapter Sara Beckman answers these questions, discusses in detail the types of variability to which flexibility responds, and cites examples of appropriate use of production flexibility.

INTRODUCTION

A new philosophy about manufacturing is sweeping industry: Manufacturers are learning to be flexible and to use flexibility, along *with* quality and productivity, as a competitive weapon in a rapidly changing and fiercely competitive marketplace. A number of factors are contributing to the heightened interest in flexibility:

- The rate of new product introduction has increased as product life cycles have shortened, particularly in the so-called high-tech industries.
- Increasing competition from foreign producers is challenging U.S. manufacturers to provide their traditionally broad product mix at lower cost. The Japanese set the stage for low-cost, high-

quality manufacturing primarily with standardized products. U.S. consumers prefer multiple-option products but want them at cost and quality levels similar to those provided by the Japanese on comparable products.

- Customization has become critical to the penetration of both foreign and domestic marketplaces. More sophisticated consumers throughout the world expect products to be tailored to their specific needs. A few years ago, computer users in China had to live with English keyboards; they can now procure computers that "speak" their native language.
- Technology is evolving at an increasingly rapid rate, requiring manufacturers to quickly assimilate new materials, as well as new process and product technologies into their organizations.

In short, change is the name of the game. And flexibility is the response that successful companies are employing to cope with it.

At the same time that the need for flexibility is growing, new technology in the form of flexible manufacturing systems (FMS), computer-controlled equipment, and robotics is being developed that allows for more actual flexibility on the production floor at lower cost. Simultaneously, information technology developments are facilitating information transfer, thus improving the overall responsiveness of the organization. While there are many mechanisms for achieving flexibility that are not technology-based, technological developments have been critical drivers in the creation of flexible organizations.

Although the coexistence of the need for flexibility and the technology to provide it is unprecedented in recent manufacturing history, the concept of flexibility as a primary dimension of manufacturing performance is not new. Wickham Skinner proposed in 1978 that flexibility, along with cost, quality, and dependability were the critical aspects of manufacturing performance. Of these, cost, quality, and dependability have all received considerable attention and are fairly well understood. Recent advances in cost accounting and capital investment justification techniques continue ongoing efforts to understand manufacturing costs. Concepts of quality control have been well established for many years, and have received additional notoriety in the form of total quality control programs. In addition, although the definition of dependability has changed some over the years from achieving monthly shipping targets to meeting customer due date requirements, it, too, is well understood.

Flexibility remains the most poorly defined and understood dimension of manufacturing performance.

Although a number of authors, primarily academic, have recently begun to address the topic of flexibility, the conflicting definitions purporting to explain why flexibility is needed and what the means are for achieving it are still confusing. This chapter attempts to reduce that confusion. We will describe the reasons an organization would want to be flexible and the importance of linking those reasons to the firm's overall strategic direction. Second, we will describe how organizations can achieve flexibility. Finally, we will discuss the trade-offs to be made in developing a flexible organization.

WHAT IS FLEXIBILITY?

In the most basic sense, *flexible* means "responsive to change; adaptable; capable of variation or modification." Alternatively, *flexibility* is defined as "the ability to respond effectively to changing circumstances." Manufacturing flexibility might then be defined as the ability of a manufacturing organization to deploy and redeploy its resources effectively in response to changing conditions. Several authors have attempted to describe flexibility. A summary of their work is provided in Table 6–1. The definitions provided range from quite narrow to very broad, from describing types of variability to describing mechanisms for achieving flexibility, and from applications to the field of economics to illustrations from management science. Collectively, they present a very confusing picture.

Because flexibility is a response to changing conditions, to variability, it seems most reasonable to begin defining it by categorizing and describing the types of variability to which a manufacturing organization must respond.

Requirements for Flexibility

There are five primary sources of variability that may require a manufacturing organization to be flexible: demand variability, supply variability, new product introduction or product variability, new process introduction or process variability, and workforce and equipment variability. As

TABLE 6–1
Definitions of Flexibility

Author	Type of Flexibility	Definition
Gerwin[a]	Mix flexibility	The processing at any one time of a mix of different parts loosely related to each other
	Parts flexibility	The addition of parts to the mix and removal of parts from the mix over time
	Routing flexibility	The dynamic assignment of parts to machines—that is, the rerouting of a given part if a machine used in its manufacture is incapacitated
	Design-change flexibility	The fast implementation of engineering design changes for a particular part
	Volume flexibility	The accomodation of shifts in volume for a given part
Buzacott[b]	State flexibility	The ability of the system to process a wide variety of parts of assemblies without intervention from outside to change the system
	Job flexibility	The ability of the system to cope with changes in the jobs to be processed in the system
	Machine flexibility	The ability of the system to cope with changes and disturbances at the machines and work stations
Browne[c]	Machine flexibility	Ease of making changes to a given set of parts
	Process flexibility	Ability to produce a given set of part types, each possible using different materials, in several ways
	Product flexibility	Ability to change over to produce a new (set of) product(s)
	Routing flexibility	Ability to handle breakdowns and to continue producing a given set of parts
	Volume flexibility	Ability to operate profitably at different production volumes
	Expansion flexibility	Capacity of building a system, and expanding it as needed, easily and modularly
	Operation flexibility	Ability to interchange the ordering of several operations for each part type
	Production flexibility	Ability to produce a large universe of part types
Jaikumar[d]	Process flexibility	Ability to reroute a part when a machine is down
	Program flexibility	Ability to run the system unattended during a shift
	Product flexibility	The total incremental value of new products that can be fabricated within the system for a defined cost of new fixtures, tools, and parts programming

[a] D. Gerwin, "The Do's and Don'ts of Computerized Manufacturing," *Harvard Business Review* 60, vol. 2 (March–April 1982). pp. 107–116.
[b] J.A. Buzacott, "The Fundamental Principles of Flexibility in Manufacturing Systems," *Proceedings of the First International Conference on Flexible Manufacturing Systems* (Brighton, UK, 1982).
[c] J. Browne, "Classification of Flexible Manufacturing Systems," *The FMS Magazine*, April 1984, pp. 114–117.
[d] R. Jaikumar, "Flexible Manufacturing Systems: A Managerial Perspective," unpublished working paper, Harvard Business School, 1984.

we describe them below, think about your organization and the extent to which it must cope with each of the types of variability.

Demand Variability

Demand variability is well-recognized as a source of disruption in manufacturing. A plethora of demand-forecasting models attests to the fact that demand variability is an important factor in manufacturing businesses. These models attempt to assess trends and seasonality in demand and to identify levels of aggregation at which forecasts can most accurately be made. Inventory planning models, of which the economic order quantity (EOQ) models are the most popularly used, were also developed in response to a need to cope with variable demands on manufacturing. The advent of just-in-time manufacturing has renewed focus on demand variability and its impact on the design of the manufacturing process.

Demand variability arises in two forms. It may be caused by the existence of a broad product mix, which may also vary over time. Automobile manufacturers typically provide a wide variety of options, thus practically ensuring that they never produce two cars in a row that are exactly alike. They must learn to cope with demand variability in the form of a broad and variable product mix.

Demand variability may also arise in the form of volume variability. Suppliers to retail sales channels often experience considerable variability in sales on a seasonal basis as their customers make holiday, beginning-of-school, and other seasonal purchases. Volume variability may be experienced by a manufacturer of a single product line, or may compound the already complex management problem of broad product mix.

Thus, breadth of product line and variation in demand volume are both sources of demand variability with which manufacturing organizations must cope. How many different products does your organization sell? Do sales vary much from month to month? If your answers are "many" and "a lot," then your organization is clearly dealing with **demand variability**.

Supply Variability

Supply variability has also been widely recognized as a source of disruption for a manufacturing operation. Models to set safety stock levels and reorder points all recognize uncertain delivery lead times and variable incoming material quality as factors. Scrap and rework rates are accounted for in setting yield factors in materials requirements planning systems. Minimizing supply variability is recognized as a basic tenet of just-in-time manufacturing.

As with demand variability, there are different dimensions of supply variability. First, the number of different parts to be procured is certainly a source of complexity similar to that introduced by the number of different products along the demand dimension. It likely implies that the organization must deal with a number of different suppliers as well.

In addition, timeliness of parts delivery and quality of parts received contribute to supply variability. Despite improved relationships with suppliers, uncertainty as to delivery and quality still remains. As the new Apple Macintosh facility was ramping up production in early 1984, it suffered two highly publicized experiences with supply uncertainty. A defective shipment of CRT's arrived from Japan that shut down the assembly facility briefly. About the same time, a shortage of memory chips also closed the facility. Despite just-in-time ties with vendors, supply was not 100 percent guaranteed. Furniture manufacturers, too, must cope with different qualities of raw materials as lumber is drawn from different locations. A sugar beet refiner deals with variability in sugar content of the raw beets fed into the refinery. Although these companies are in very different industries, they must all deal with uncertainty about quality and timeliness of material delivery.

The introduction of new materials is yet another source of supply variability. Airframe manufacturers are still learning to fabricate composite (rather than aluminum) parts. Semiconductor manufacturers are accommodating gallium arsenide and ceramic substrates.

In all, number of different parts sourced, uncertainty as to timing and quality of delivery, and new materials all contribute to supply uncertainty in a manufacturing business. Does your company purchase many different component parts? Are those parts ever delivered late? Are they of variable quality? Have you introduced any new materials to your manufacturing process lately? If you answered "yes" to any or all of these questions, then your organization must be dealing with **supply variability**.

Product Variability
Product variability can range from the *introduction of completely new products* to the daily changes implemented on existing products.

The rapid pace of technological evolution is causing industries to experience shorter product life cycles and higher product turnover. From 1964 to 1976, for example, IBM introduced only two families of mainframe computers. Between 1976 and 1980 new computer families were introduced at the rate of one per year. New product introduction is therefore imposing a major requirement for flexibility on manufacturing businesses.

Closely related to new product introduction is product change. Your company would be highly unusual if it didn't experience numerous "engineering change orders" on its products throughout their lives. Whether these changes are initiated by the customer, by engineering, or in production itself, they are clearly sources of variability with which the organization must cope.

Thus, new product introductions and changes to existing products constitute what we call **product variability** for a manufacturing business. Does your organization experience product variability?

Process Variability

Process variability comes in two forms: introduction of new process technology and introduction of new process management techniques.

Installation of new process technology is often closely related to the development and introduction of new products. Semiconductors, for example, have become smaller as the technology used to produce them has become capable of accurately embedding more functionality in smaller spaces on these devices. In this case, by necessity, introduction of product and process technologies go hand in hand.

In the machining industry, on the other hand, computer-controlled equipment and flexible manufacturing systems have significantly changed processes for manufacturing complex machined parts by reducing setup times and improving accuracy. While use of the new equipment is not necessarily required to build new products, the performance of the process is greatly improved.

In all, new process technologies such as these—whether developed internally, procured from equipment vendors, or copied from competitors—create complexity for manufacturers.

In addition to the changes effected by the introduction of new process *technologies*, there is change associated with the introduction of new process *management* techniques. In the recent past, materials requirements planning (MRP) caused hundreds of companies to make fundamental changes in the management of their production processes. More recently, just-in-time (JIT), total quality commitment (TQC), design for manufacturability (DFM), and computer-integrated manufacturing (CIM) have all had an impact on the way that manufacturing organizations do business. These changes in process management can be, but need not necessarily be, accompanied by the introduction of new process technology as well.

Introduction of process change, whether in technology or in process management, imposes a degree of variability on the manufacturing organization. Do process technologies evolve rapidly in your industry? Do you have to develop new processes for the new products you introduce? Are you keeping up with the latest trends in new process management? If so, then your organization is dealing with what we call **process variability**.

Workforce and Equipment Variability
Finally, both labor and equipment employed in the production process are sources of variability. Labor turnover, absenteeism, and efficiency are causes of uncertainty in planning production. Economic conditions, the location of the manufacturing facility, the personnel policies of the company, and the treatment of the workforce influence the degree to which labor uncertainties are experienced.

In terms of consistent production of quality product and machine uptime, variations in equipment reliability can also create a source of uncertainty that has been well-recognized in management models for doing job shop and preventive maintenance scheduling. Machines that wander out of tolerance as they operate, or that require frequent repair, cause an increased requirement for flexibility in a production operation.

Thus, both **workforce and equipment variability** constitute the fifth type of variability to which manufacturing organizations must be responsive.

Summary—Five Types of Variability

We have argued that flexibility in manufacturing must be developed in response to some well-defined strategic need. In this section, strategic needs have described in terms of variability in the variety of forms shown in Table 6–2. A clear understanding of the types of variability your

TABLE 6–2
Uncertainty and Complexity Dimensions of Variability

Type of Variability	Uncertainty Forms	Complexity Forms
Demand variability	Fluctuation in: Product mix Volume	Product mix
Supply variability	Changing delivery: Quality Timeliness	Number of parts New materials
Product variability		New products Changes to existing products
Process variability		Changes in process: Technology Management
Workforce and equipment variability	Labor: Absenteeism Turnover Equipment downtime	

organization chooses to contend with to provide a competitive advantage is critical. Note that the source of variability need not necessarily always be external to the business itself. Once the workers and equipment sets are chosen and either hired or installed, workforce and equipment variability are largely internally generated. Demand variability, on the other hand, is more likely to originate outside the organization with the customer. Product variability lies somewhere in between. Hopefully, product changes are driven largely by customer demand, but they may also be caused by internal reasons. Manufacturing may wish to have a product redesigned to make it more manufacturable, or marketing may ask that an additional feature be added. The organization must decide with how much and what type of variability it wants to contend, whether internally or externally derived.

Let's examine a few examples of organizations that have developed flexibility in response to a perceived strategic need.

IBM's "Quick Turnaround Time" integrated circuit fabrication facility in East Fishkill, New York is capable of building prototype ICs routinely in eighteen days and, on an expedite basis, within three days. Most competitors require at least a six-week turnaround time on new designs. Strategically, this allows IBM to develop and introduce new products rapidly, as well as to economically produce custom ICs for pilot runs and small volume sales. IBM is choosing to deal competitively with what we called product variability.

Allen-Bradley spent $15 million over a period of two years to construct a fully automated motor-starter assembly plant in Milwaukee, Wisconsin that today builds over 700 different versions of motor starters in lot sizes as small as one. Orders received at the plant on one day are manufactured, tested, packaged, and shipped the next. At a time when most of its competitors were moving to the Far East to obtain low-cost labor for manual product assembly, Allen-Bradley has been able to obtain a considerable edge. In this case, Allen-Bradley is seeking competitive advantage in its ability to effectively and efficiently manage demand variability.

Even in the automobile industry, where long production runs on fixed-setup assembly lines have been a long-standing tradition, flexibility is being strongly considered in plant design and redesign decisions. General Motors plans to invest $52 million in an automated, highly flexible manufacturing plant in Saginaw, Michigan, to machine and assemble a family of axles for several different models of cars. The plant is a direct response to the shortening of product life cycles in the auto industry. It provides manufacturing capabilities that accommodate themselves to rapid design changes and product customization requirements.

For lower volume automobile manufacturers (e.g., Saab, BMW, and Alfa Romeo) manufacturing flexibility provides an opportunity to compete with the high-volume, low-cost producers. BMW pioneered sophisticated techniques for varying product mix randomly down common production lines, allowing increased resource use despite low unit-volume production. Austin Rover made extensive use of computer-aided engineering to reduce the new product introduction cycle from five years to three and a half years, permitting them to bring new designs to market more quickly. Alfa Romeo is piloting a flexible final assembly process using the Volvo "group assembly" production model to achieve benefits similar to those sought by BMW.

All of these companies recognize manufacturing flexibility as a critical aspect of being competitive in their marketplace and are organizing their manufacturing resources to achieve it.

The possibility that your organization may have to cope with all five types of variability at one time is a daunting prospect. Yet, most of our organizations do deal with them all. Whether we can deal with all of them well is a good question, and one worth considering in the development of the strategic direction of the business. Let's look at the types of mechanisms that are used by companies to make themselves more flexible along each of the dimensions we've described. Where conflicts among these mechanisms arise, potential for not achieving the strategic objectives of the firm exists.

CREATING PRODUCTION FLEXIBILITY

Traditionally, manufacturing organizations appeared flexible to the outside world by carrying inventory. In order to protect the production organization from the vagaries of customer demand, they carried finished-goods inventory. To obviate the disastrous impact of supply variability, they carried raw-materials inventories. And in case of variability in the process itself, in either equipment or labor, they carried work-in-process inventory. Thus, as shown in Figure 6–1, the production organization was well protected from the variability of the outside world, as well as from internally generated variability, and the organization as a whole appeared capable of responding to variability in demand, supply, and workforce and equipment.

Total quality commitment and just-in-time philosophies have forced a rethinking of the use of inventory as protection for the production organization. Many manufacturing organizations have steadily reduced their inventory levels to the bare minimum and expect the production organization to be able to respond directly to the variabilities it faces. Other organizations are struggling to eliminate that last layer of inventory, perhaps because they have not yet provided production with the capabilities it needs to respond to change in the absence of inventory.

In addition to the philosophical changes encouraged by TQC and JIT, new product and process introductions are becoming more and more important. Inventory not only rarely protects production from changes in product or processes; it often becomes a deterrent to accommodating the required changes. Does your organization ever wait to implement an engineering change to a product because it wants to use up existing

FIGURE 6–1
Inventory Buffers Protecting Production from Supply, Demand, and
Workforce and Equipment Variability

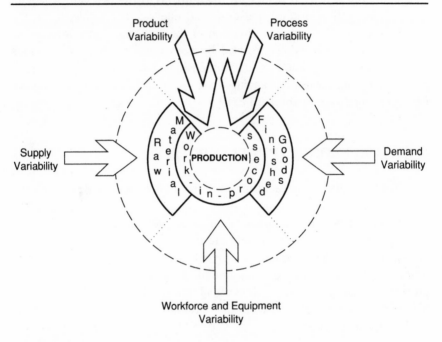

Workforce and Equipment
Variability

inventory first? The production organization is being forced to invent
and implement a new set of coping mechanisms for dealing with all of
the different types of variability.

The New Flexible Production Organization

Much has been done in production organizations to improve their
flexibility. New, more flexible automation is at the heart of much of
this change. Workforce flexibility is absolutely critical. And some cases
require carrying some form of excess capacity—in operations research
terms, "slack resources."

Automation

Allen-Bradley's motor-starter plant is one of the best publicized examples of the use of automation to achieve flexibility, but there are others.

Motorola's "Bandit" facility in Boca Raton, Florida, uses automation to produce a wide variety of pagers on demand. Benetton, the Italian sportswear company, has not only automated the knitting process for producing clothing, but has integrated the entire supply chain—from raw materials through the retail outlets that sell its wares. By doing so, it is able to respond rapidly to changing demands all over the world.

There are several different aspects to automating for flexibility. Numerical control and other equipment programmed for rapid setup allow for quick modification in response to product mix variability, new product introduction, and product change. Automatic tool-changing ability or the ability to keep different tools online similarly accommodate changing product mix and often new products as well. Automated material-handling equipment and sophisticated part-loading devices can expedite delivery of materials to the production system. Finally, investment in multipurpose equipment provides flexibility as well. Thus, there are a number of aspects of automation to keep in mind if flexibility is an objective of your organization.

Another important aspect of most automated environments is automated information management, often referred to as CIM. Provision of just-in-time information to the people or equipment performing the production process can have a critical role to play in providing ultimate flexibility. Computer aided design/computer aided manufacturing (CAD/CAM) links allow for rapid translation of new product designs or design changes from the drawing board to the production floor. Real-time scheduling systems allow products to be rerouted around bottleneck activities. Data collection for feedback and control purposes is equally important. Having sensors and computer controls for detection and handling of unanticipated problems (poor material quality, equipment breakdowns, etc.) allow for rapid response to these forms of variability.

Automation, then, whether of the physical process or of the associated material flows can be an important means of achieving flexibility in the production organization. In the quest to implement the latest process technologies, however, the human aspects of the production system must not be neglected.

Workforce Flexibility
Development of a flexible workforce must go hand in hand with implementation of flexible automation if automation is a chosen solution. In

other cases, a flexible workforce alone may provide the needed degree of flexibility. There are a number of ways in which production organizations are improving the flexibility of their workforces.

A large number of companies are attempting to integrate shop floor workers more fully into the production process. In doing so, companies typically provide extensive training to the workers in statistical process control techniques so that they can attempt to stabilize the process, even in the face of constantly changing inputs or required outputs. Further, the company may cross-train workers in multiple tasks. Cross-training increases production flexibility in several ways: It familiarizes employees with all activities throughout the production process, thus giving them a better picture of the whole and how their job affects overall output. It allows for efficient redeployment of the workforce as demand changes. Absent employees' jobs can be more easily covered and workstations can be quickly restructured without considerable disruption or need for extensive retraining.

Cross-training to achieve flexibility is not, however, restricted to manufacturing industries. Lechmere, Inc., a 27-store retail chain owned by Dayton Hudson had used the traditional approach to covering variability in customer traffic through their stores by hiring a large number of part-time workers who could be scheduled to cover peak demand requirements. Around Sarasota, Florida, where unemployment rates run as low as 4 percent, this approach was not feasible. Instead, the Sarasota store offers its employees raises based upon the number of jobs they learn to perform. Having motivated a highly cross-trained workforce, they can accommodate changes in customer demand by rapidly redeploying the workforce as needed. Cross-training was an important element of Lechmere's ability to compete in the Sarasota environment.

Another tool used by many companies is the concept of flex-force. Employing a flex-force entails qualifying a list of people from the local community that are interested in periodic, part-time work—people the company can call during periods of need. These people do not become permanent parts of the company payroll, so they do not cost the company anything when they are not working, but they are an available, trained resource that can be tapped when necessary. They can be used either to respond to variability in demand volume, or to fill in capability gaps caused by other types of variability. For example, demand for printed circuit board assemblies may be such that an unusually high amount of hand load work is required in a given month. Availability of a ready, trained workforce may be very useful in responding to such a demand.

A flexible workforce, achieved through increased employee involvement, workforce cross-training, or development of a well-trained flex-force, is likely to be a critical aspect of achieving production flexibility.

Other Mechanisms

There are a number of other ways in which production can become more flexible that do not necessarily require investment in equipment or a flexible workforce. In the quest for just-in-time manufacturing, many organizations have significantly reduced setup times. They have done so through parts kitting, equipment dedication, production sequencing, and other such techniques. Lower setup times improve production's responsiveness to various types of variability.

The production organization may also choose to carry excess capacity (slack) so that it can be made available during periods of peak demand or so that there are extra resources to be deployed as product mix varies. The company may maintain excess capacity internally in the form of employees or equipment. Or it may choose to develop strong relationships with subcontractors to provide services in time of high demand. In particular, the company may contract capacity to supplement its bottleneck operations. A sheetmetal production facility, for example, that chronically runs short of capacity at its punch presses may occasionally choose to subcontract some of its punching requirements.

Summary

There are a number of techniques that production can employ to become more flexible in response to any of the five types of variability, including:

- Employ flexible automation.
- Integrate manufacturing information management.
- Cross-train the workforce.
- Employ a "flex-force".
- Reduce setup times.
- Maintain excess capacity, internal or external.

Most organizations use a combination of these techniques to improve production's overall flexibility.

FLEXIBILITY IS NOT PRODUCTION'S JOB ALONE

Thus far we have focused our discussion on the activities that production can undertake to improve the flexibility of a manufacturing business. It is critical, however, that all functions in a manufacturing business be involved in the achievement of flexibility. In particular, marketing and R&D have important roles to play. Production cannot and should not be asked to shoulder the entire burden of providing flexibility by itself.

Marketing's Role in Creating a Flexible Organization

The marketing organization plays an important role in managing the interface between the customer and the manufacturing organization. In assuming this role, it can have a significant effect on the amount of demand variability that eventually hits the production operation. There are a couple of ways that marketing can effectively smooth demand before it reaches production, neither of which should significantly affect customer perception of responsiveness from the organization.

First, marketing is responsible for planning and executing promotional activities that have direct impact on demand. By planning these activities to coincide with periods of low demand, marketing can effectively smooth production's build schedule. Postholiday sales are a good example of attempts to smooth demand (and thereby revenue). While it is unusual to actually find a marketing organization that thinks about manufacturing in its planning of promotional events, it doesn't seem unreasonable to expect that they could. They are accustomed, after all, to hosting "fire sales" to rid the organization of excess inventory *after* production has built it. Why not plan ahead?

Marketing also plays a critical role in the definition and positioning of the product. If it performs this exercise creatively, it can significantly reduce the impact of a broad product mix on production. Hewlett-Packard's Signal Analysis Division, for example, recently introduced a line of new instrumentation products that consist of multiple modules integrated in a single rack. Each module has different functionality. Together, a collection of modules creates an instrument that meets a certain set of customer specifications. By mixing and matching the modules, the specific requirements of any given customer can be met; production has only to build the small number of modules that make up the total instrument.

The customers perceive that they are receiving customized products, while production is dealing with a minimal product mix. By defining product features in small packages that can be combined as required to meet customer requirements, marketing was able to minimize the impact on production of multiple customer demands.

Thus, marketing can have a significant impact on the variability and on the perceived flexibility of the company. It can either reduce the absolute variability experienced by the organization through promotional events, or it can support definition of a broad product mix through modularity in product design. In the latter exercise, it must work closely with the R&D organization as well as with manufacturing.

R&D's Role in Achieving Flexibility

As rapid new product introduction becomes increasingly important to successful competitive positioning, R&D's role in providing flexibility will grow. Most often, the way in which R&D can make a contribution to the overall flexibility of the organization is through application of a set of standard design rules and use of a standard set of components in their designs. Both are critically important to the ability of the entire organization to produce and deliver new products to the marketplace in a timely fashion.

It is becoming quite popular to embed a set of design rules in the computer-aided design packages used by product engineers. These rules direct the designer to use product characteristics that best match the production process currently in place. They may, for example, describe part-spacing requirements on a printed circuit board assembly or optimal fold angles on a sheetmetal box. In any case, they are intended to reflect the cost trade-offs inherent in the production process in which the product will be made.

Some systems even go so far as to present the designer with explicit cost information. At Hewlett-Packard's Roseville Networks Division, an expert system called the Manufacturing Knowledge Expert (MAKE) provides cost estimates to printed circuit board assembly designers that allow them to make trade-offs during the design process. They can immediately see, for example, the impact of including a part in their design that will have to be hand-loaded rather than automatically inserted. Selecting the low-cost solution will probably result in the fastest throughput time as well, as it likely employs the most available processes.

Just as critical to flexible operations is the use of a common set of parts in a product design. Many manufacturing organizations have undertaken programs to significantly reduce the number of parts used in their products. IBM, for example, reports a 25 to 30 percent reduction in the number of parts used to produce their products. Others report similar reductions. Using parts from a common and existing set of preferred parts provides an obvious advantage for achieving flexibility. Production doesn't need to qualify a new set of vendors on a new set of parts; it can use parts it already procures to build the new product. New products can be built and tested much more quickly. In addition, fewer parts implies the possibility of fewer setups, thus reducing cycle time and improving the ability to respond to broad product mix.

In sum, R&D has an important role to play in the achievement of manufacturing flexibility through adherence to a strict set of design rules and selection from a limited set of preferred design parts.

ORGANIZATION STRUCTURE AND FLEXIBILITY

We have described the individual roles that production, marketing, and R&D have to play in the achievement of flexibility, each of which is important and unique. But they must all ultimately play together to provide flexibility, and their ability to do so is a function of the design of the organization.

In his "information processing view of an organization," Jay Galbraith provides a nice description of the effect of variability on an organization and of the design choices that an organization makes to respond to variability. [1] An organization that deals with little variability from any source can be run largely through the application of rules and procedures. Any exceptions can be processed by passing them up the hierarchy. As the organization faces increasing variability, however, the number of exceptions to be handled becomes overwhelming and the efficiency of the hierarchical organization begins to break down. In some cases, replacement of the rules and procedures with more broadly based guidelines helps. Ultimately, however, the organization will likely have to select a new structure that is more suited to the variable environment in which it operates. It has a couple of options.

First, the organization may choose to improve its communications ability both laterally and vertically. The more effectively and efficiently information can be communicated either across the organization or up and down the hierarchy of command, the more rapidly the organization can deal with change. The advent of computer-integrated manufacturing (CIM) clearly addresses communication requirements. By managing common manufacturing information and ensuring that the most current information is available as needed, CIM improves the organization's responsiveness to variability. Tracy O'Rourke, CEO of Allen-Bradley, claims that the most difficult part of the implementation of their automated motor-starter plant was not the automation but the information management system that overlays the automated factory. Clearly, well-executed information management is what allows the factory to build 700 different versions of the product on demand.

The organization may also choose to restructure into what Galbraith calls "self-contained tasks." By this he means structuring the organization into subunits that are focused on well-defined outputs. Further, each subunit is provided with all of the resources it needs to provide its output. Hewlett- Packard's (HP) product-focused organization is a good example. Rather than complicate a large organization with handling the rapidly expanding product line of instrumentation and computational products, HP spun off individual product divisions that had full capability and responsibility for a limited set of products. Each division contained marketing, manufacturing, R&D, finance, and personnel functions to support their product charter. Only the sales organization remained centralized. Each division had to deal only with the variability imposed by the specific marketplace for its products, a much simplified set of the total HP demand.

Another form of self-contained tasks commonly found in practice is the application of group technology to form-machining centers or cells. In this case, parts with common characteristics or requiring a common set of equipment or tooling are grouped together. All of the production equipment needed for their manufacture is co-located in a cell, many of which may make up the total production floor. These cells then concentrate on the production of a limited number of similar types of parts and do not experience the full variability of the factory.

It is important to understand that these concepts can be applied at various levels in the manufacturing organization. While sweeping organizational changes must be instigated by senior management there are similar applications of these concepts that can be made at middle- and lower-management levels as well. Application of just-in-time concepts, even in small parts of a factory, effectively tightens communication linkages between individuals. (There is no longer a wall of inventory preventing communication between workstations.) Computerized information management within a workcell is another possibility.

Quality circles, which are today evolving to high performance work systems, are yet another form of self-contained task. They effectively serve to bring together all individuals involved in the creation of a specific output to work on the problems associated with producing that output in an efficient, effective manner. In doing so, they draw together in one place all of the resources associated with production of that output. By limiting their focus to a small number of outputs, they effectively limit the variability with which they must deal.

Organizational structure is critical to the achievement of manufacturing flexibility. Through improved communication linkages, either vertical or horizontal, and the creation of self-contained tasks, an organization's ability to handle variability, to be flexible, can be greatly improved.

THE FLEXIBILITY MODEL

We have defined the five types of variability for which an organization may wish to develop flexibility: demand, supply, product, process, and workforce and equipment variability. Further, we have described the roles not only of the production function, but of the marketing and R&D functions as well in responding flexibly to demand for change. Finally, we have discussed the importance of organizational structure in the organization's ability to be flexible. We have a model of manufacturing flexibility (Figure 6–2).

There are many trade-offs to be made. First, the organization must understand which types of variability are to be dealt with in achieving its competitive advantage. Will it attempt rapid and frequent new product introduction? Or will it produce a limited product line for which volume variability is an issue? Is it operating in a business (e.g., sugar refining or logging) in which supply variability is a significant factor? Or is process variability the name of the game as it is in the semi-conductor industry? Whatever its primary mode, it is important that your company

FIGURE 6–2
R&D and Marketing Roles in Managing Variability

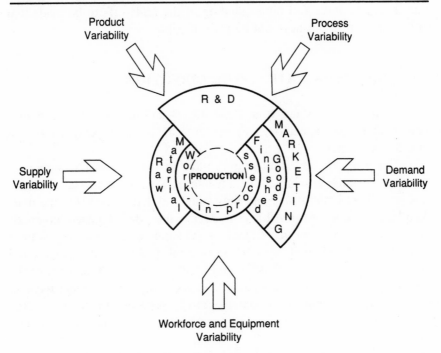

understand what types of variability it is dealing with, as they each may require different types of flexibility on the part of the organization. Ultimately, most of the trade-offs occur in production, where all variability not managed elsewhere comes to rest. There are many questions to be asked. Will the "flexible" equipment that is being purchased to build the existing broad product mix also accommodate numerous new product introductions? Or would it be better to employ a more manual process under the assumption that *people* are easier to set up for new products than equipment is? Will limiting part count in an attempt to minimize the impact of broad product mix on production cause it to be inflexible in new-product introduction efforts? The production function has much to gain from an understanding of how it will achieve flexibility along all of the required dimensions.

The astute manufacturing manager will make clear both the types of variability to which his or her organization must respond and the role that the marketing and R&D organizations are expected to play in creating the ultimate flexible organization. It is only through cooperation among the functions and a full understanding of the trade-offs to be made that a manufacturing business can be truly flexible.

FLEXIBILITY IS A STRATEGIC DIRECTION

Now that we understand what flexibility is and how to achieve it in a manufacturing organization, let's return to the strategic perspective with which we started.

For years, U.S. manufacturers have focused on cost reduction as a primary source of competitive advantage. Through increased machine use and direct labor utilization, they sought to deliver goods to the marketplace at the lowest possible cost. Meanwhile, their Japanese competitors, listening to the words of Deming and Juran, selected quality as their primary focus. Through application of total quality control (TQC) and statistical process control (SPC) along with quality circle and employee involvement structures, they pursued productivity improvement *through* quality improvement. U.S. manufacturers persisted in believing that quality could be achieved only at great cost and, as a result, continued to lag behind in both productivity and quality.

What we did not understand during those times of great focus on single dimensions (i.e., cost or quality) of manufacturing performance is that we could successfully pursue more than one dimension simultaneously. Why? The dimensions are not orthogonal, but rather may be fully complementary. Quality and cost were eventually seen as complementary: Pursuing quality made cost reduction, or productivity improvement, possible. There is a growing belief that flexibility, too, can be an objective complementary to cost.

The Changing Product / Process Matrix

Hayes and Wheelwright[2] popularized the concept of the product/process matrix shown in Figure 6–3. The best place to position a manufacturing organization, they argued, was on the diagonal of the matrix where process capabilities are best matched to product requirements. The existence today of flexible equipment changes this picture (see Figure 6–4). By employing flexible automation along with the techniques described

FIGURE 6–3
Hayes-Wheelwright Product/Process Matrix

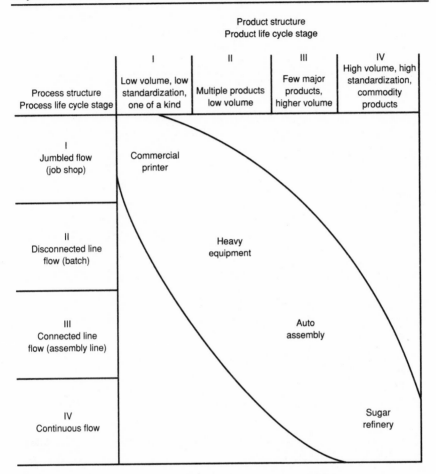

in this chapter for achieving flexibility, organizations are able to afford low cost factories that can serve multiple types of product requirements. The new term being used to describe this picture is *economies of scope*. In contrast to *economies of scale* where cost advantage was gained through production of great volumes of few items, economies of scope implies that similar gains can be achieved by producing many different products whose individual volumes add up to large total plant volume. Flexibility is critical to achieving economies of scope.

FIGURE 6–4
Product/Process Matrix Historical Evolution

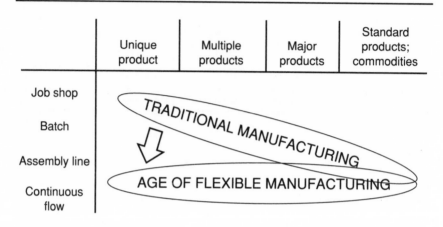

Flexibility and the Japanese

Although Japanese manufacturers appear to be rapidly evolving new capabilities for manufacturing flexibility, they have traditionally not been as flexible as U.S. manufacturers. A comparison of the two approaches sheds some light on the strategic role of flexibility in market positioning.

The Japanese focus considerable attention on maintaining flexibility on the shop floor to allow rapid response to manufacturing disruptions, such as machine breakdowns and quality problems (the category we have labeled workforce and equipment variability). To do so, they use cross-trained labor, investment in redundant equipment (excess capacity), U-shaped rather than L-shaped layouts, and other such techniques. Historically, U.S. manufacturers have focused more on resource utilization and have therefore tended more often to dedicate labor and equipment to specific tasks. Furthermore, the Japanese choose local suppliers, and through JIT materials-management techniques are able to maintain flexibility to respond to engineering design changes, production schedule changes quality problems, and other related changes without huge inventory write-offs.

There is, however, an apparent limit to the amount of flexibility that is considered acceptable to ask of a Japanese manufacturing organization. Production schedules are often frozen at a certain point, and Japanese companies often reduce product-line breadth or increase customer minimum-order sizes rather than allow significant variation in demands placed on the manufacturing organization. U.S. manufacturers have created markets through product differentiation and have been much more responsive to the demands of the marketplace because of their ability to customize products and maintain multiple-product options.

Support of lifetime employment policies in Japan may limit their flexibility in handling demand volume fluctuations through workforce sizing. Unionization of the workforce and existence of lengthy work-rule lists may provide a similar limitation in the United States; however, Japanese attitudes towards flexible use of the workforce are currently being debated by the United Auto Workers. The NUMMI (New United Motors Manufacturing, Inc.) concept—an American-flavored Japanese labor-management system—requires assembly line workers to go from performing individual, narrowly defined tasks to working as a part of a team. Application of such a concept gives management greater flexibility in running the plant but requires workers to have a broader skill base.

Finally, there are indications that Japanese organizations are quite rigid in handling new product introductions. In order to develop and introduce new products successfully, they often establish separate new product development project teams that operate independently of the rest of the organization (a form of self-contained task). IBM used a similar technique for the development and introduction of the PC Jr.

Thus, relative to the United States, the Japanese emphasize shop floor flexibility in response to machine and quality disruption but limit their responsiveness to marketplace demands for customization and broad product mix. They do not view workforce size as a source of flexibility, but do extensive cross-training so the workforce in place can be used flexibly. They apparently have difficulty innovating and introducing new products efficiently within their existing structures. Such contrasts suggest that there are in fact different strategic approaches to flexibility.

CONCLUSION

Flexibility is an important strategic choice to be understood and made by manufacturing businesses today. An astute manager will first recognize the different reasons for having flexibility—responsiveness to demand, supply, product, process, or workforce and equipment variability—and then put in place the mechanisms appropriate to achieving the desired type and level of flexibility. When used judiciously as a complement to cost and quality objectives, flexibility is potentially the new basis for competition in the next decade.

NOTES

1. Jay Galbraith, "Organizational Design: An Information Processing View," *Interfaces*, May 1974, pp. 28–30.
2. Robert H. Hayes and Steven C. Wheelwright, *Restoring Our Competitive Edge: Competing Through Manufacturing*, (New York: John Wiley & Sons, 1984) pp. 208–223.

CHAPTER 7

STRATEGIC LOGISTICS—A TOTAL COMPANY FOCUS

Hal Mather

Editor's note

What is strategic logistics? In this chapter, Hal Mather answers this question by describing various functional-area contributions to improved performance, from design engineering through sales and production. The model of a utopian manufacturing company is explored, using several innovative tactics to achieve the ideal: the design architecture statement, creation of logistically friendly designs, and realistic application of cost accounting.

In a utopian manufacturing company, vendors deliver the right materials at the right time, they are processed effectively in the plant, and quality products are delivered to customers when needed. This total-product flow is the mission statement of good logistics. It encompasses all activities between vendors and customers, from Research and Development (R&D) to after-sales service; consequently, every business function can either help or hinder achievement of the logistics mission.

Of course, the logistics objective must be accomplished profitably. There should be a differential between buying and processing costs on one hand and selling price on the other. Profits are not necessarily short-term objectives; we are quite prepared to lose money for a while if in the long run we make good profits. But effective logistics are a perennial concern: We always want to improve performance in this critical area. High customer satisfaction, excellent cash flow from reduced inventories, and flexibility to respond to changing market conditions are just some of the benefits of improving logistics performance.

YOUR P:D RATIO

The term P:D ratio comes from *Study of the Toyota Production System* by Shigeo Shingo. The *P:D ratio* is shown in Figure 7–1. *P* is the length of time it takes from ordering raw materials until they are received, processed, and shipped to a customer. This is variously called the stacked, cumulative, or aggregate lead time for a product. *D* is the customer's lead time, the time from when a customer orders a product until it is shipped.

  D can be defined in three ways:

1. D is what the company quotes as lead time to the customer.
2. D is what the customer wishes it were.
3. D is what would give the company a competitive edge if product could be reliably delivered in that length of time.

FIGURE 7–1
P:D Ratio

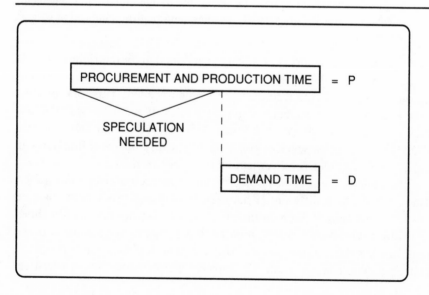

These three Ds could all be the same, or all different, depending on the product and marketplace. It does not matter which one we visualize for the remainder of this discussion. However, the strategic implications of the third definition should be obvious.

THE PLANNING DILEMMA

For most companies, P exceeds D. This leads to the planning dilemma shown in Figure 7–2. Forecasts are made for the portion of the P time not covered by customer orders. They are put into an output plan for the business called a master production schedule. This MPS is exploded through the bills of materials in a material-requirements planning run. Purchase and manufacturing orders are placed that result in intermedi-

FIGURE 7–2
The Planning Dilemma

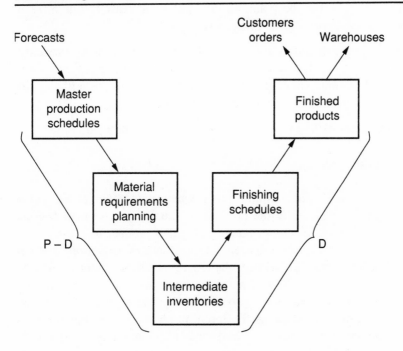

ate inventories. These could be raw materials, components, and partly finished or even finished products.

At this point, a customer's order or a warehouse replenishment order is received. These intermediate inventories are then finished off to suit the specifics on the order. The finishing-off can be as simple as pick, pack, and ship or could involve significant unique production to suit special customer needs.

The dilemma arises because forecasts are notoriously wrong. The wrong things end up in the intermediate inventories. The worst of all worlds is now apparent—huge inventories and poor customer service. The factory tries to close the gap between what was forecasted and what was actually bought by scrambling to crisis-expedite vendors. Costs now escalate with the problems of wrong inventories and poor customer service.

SOLUTIONS TO THE PLANNING DILEMMA

The choices to solve the planning dilemma or at least to minimize its effects fall into two general categories.

1. Improve the plan. In other words, make the left-hand, predictive side of the dilemma closer to the reality of the righthand side.
2. Add contingency. Buffer the inherent error in the lefthand side from the reality of the righthand side.

Several choices exist within each category; each is defined below.

Improve the Plan

1. Reduce P time. Utopia exists when P is equal to the actual processing time of the product. P equal to or less than D could also be a good interim target. Detailed product forecasts would now no longer be required. But in fact any reduction in P will be beneficial as it reduces the risk caused by erroneous forecasts.
2. Force a match. Tell or persuade your customers to buy what is forecasted. One way is to extend D time to equal P. Another is through promotions. Different product availabilities for some items versus others is a third idea. The caution here is that these

ideas must all fit within the competitive environment to avoid alienating customers.

3. Simplify the product line. It is obvious that the larger the product line, the more difficult it will be to forecast every item well. Simplification can be accomplished by truncating the line. Another method is design of multiple-application products. Utopia would feature a single product with infinite application possibilities.

4. Standardize. This means to standardize all ingredients that must be bought, and processes that must be performed *prior* to a customer's order being received. Utopia would have a single raw material from which an infinite variety of products are made by one process within the D time. A good example of this is the clay used to make crockery.

5. Forecast more accurately. Many companies do not put enough emphasis on the forecasting process, not realizing its importance to their success. It's also true that many sales policies and programs disturb underlying sales patterns, making accurate forecasting an impossibility. Eliminating these volatility amplifiers can result in significant reductions in forecast error.

Add Contingency

1. Safety stocks. If the forecast error between the left side of the dilemma chart and the right is reasonably predictable, then safety stocks can allow the forecast to be wrong while the customer is still satisfied.

2. Excess capacity. Capacity that can be turned on or off quickly to react to forecast errors can compensate for the planning dilemma. This solution requires safety stocks of raw materials.

The Strategic Issues

The seven solutions listed above can be broken into operational and strategic considerations. The strategic ones are:

1. Simplify the product line.
2. Standardize.

3. Forecast more accurately. The strategic concerns here apply particularly to sales policies and programs that cause demand instability.

Each of these stand-alone issues will be tackled in separate segments. In fact, considerable synergy will result from an attack on all three simultaneously.

PRODUCT VARIETY: FRIEND OR FOE?

The size of the product portfolio can be an emotional issue. Sales personnel want an infinite variety of products to serve every customer's needs and to build sales volume. Designers are happy to oblige with new products. They even aid and abet the sales people with new ideas and technologies. Factory people reject variety as unnecessary and very costly. They would like one product with infinite sales volume in order to really make money by honing the manufacturing machine. Accounting is no help in resolving the cost and profitability considerations of variety. They retreat into saying, "Volume spreads the fixed costs, so volume is good."

It is obvious that an infinite variety of products is too many and that one is too few. The key question is, where between these two extremes should you be to maximize return on investment?

Let's quantify the variety issue so that we can make strategic decisions about this key variable. I am not for or against variety; it is simply a business parameter. One of my clients, a manufacturer of building products, has used a variety strategy to become the country's largest and most profitable manufacturer of his product line. But his question now is, when will this strategy go too far? When will variety become a negative influence?

Another client, a manufacturer of process control equipment, has allowed variety to get so big he cannot afford to defend all market areas against niche competitors. They are slowly nibbling away at his core volume business. His concern now is how to manage product-line variety and stop encroachment by competitors into his business.

These two examples should show you why product variety (as well as other types of variety, such as customers, vendors, processes, and ingredients) are strategic issues to be managed.

Financial Implications of Variety

The issue of variety can only be addressed through communication among all business functions. As this issue is not a black-and-white discussion, it is most critical for everyone to understand the key concepts. Choices to best suit the business—either to limit variety and try to lower costs or to build variety at the expense of profits—can now be made from the limited possibilities available.

We are going to test the relationship between variety and money, using Figure 7–3 as the guide. We take each of the five factors impacted by the amount of product-line variety—revenue, profit contribution,

FIGURE 7–3
Effects of Product-Line Variety on Five Key Cost Factors

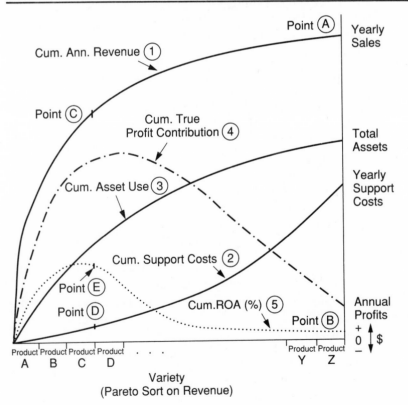

Variety
(Pareto Sort on Revenue)

asset use, support costs, and return on assets (ROA)—and plot them to demonstrate the contrasts. The plot demonstrates the principle that, as product variety increases, so do support costs.

To explain the curves, let's start with the axes. The right vertical axis is money. It starts at zero and ends at annual sales revenues. This axis can extend below the horizontal axis: Above is positive; below is negative.

The horizontal axis is variety. We arbitrarily pick the variety of end products made to define this axis in the discussion. We could just as easily have chosen one of the other varieties, such as customers or ingredients. Either of these might be a better definition for your particular business. It will be easy to make the conversion from end-product variety to key variety variable after the discussion.

The end products are arranged along the horizontal axis with the one generating most annual sales (Product A) on the far left; the one generating least (Product Z) is on the far right. Those in between are organized in descending annual sales. This is a Pareto or ABC sequence.

All curves shown in Figure 7–3 represent *cumulative* performance. For example, point C on the cumulative annual revenue curve is the combined contributions to revenue of products A through C. Moving from left to right on the horizontal axis, therefore, increases the level of variety by adding more types of increasingly specialized products.

Curve 1 is a picture of cumulative annual sales from the product variety. A typical curve shows 80 percent of the revenues come from 20 percent of the product line (point C, for example). The curve for your business could be steeper than this (90/10) or shallower (70/30). It is an easy curve to draw for a business.

Before looking at Curve 2 we need to define support costs. These can be shown in a simple calculation. Take yearly sales, subtract profits, subtract unburdened direct labor and direct material. The remainder is support costs. It includes factory supervision, scrap, rework, sales expense, accounting, industrial engineering, design engineering, administrative costs and so on. This is not a typical accounting classification. It must be engineered from accounting figures.

The key question is how to distribute annual costs over the product line. Which products cause high support costs, which ones small? A typical distribution is shown as Curve 2. The high-volume, limited-variety products incur low support costs (product D). The low-volume, high-variety products require lots of support, thereby causing the yearly support costs to rise sharply in the high-product variety range.

Some interesting work called *transactional costing* is proving the support cost distribution to be correct. In essence, transactional costing says that most support costs arbitrarily allocated to products today or classified as fixed or period costs are incurred to handle transactions. A transaction occurs when an item or data is moved, gathered, picked, ordered, entered, purchased, inventoried, sent somewhere, inspected, or stored. Low-volume, high-variety products generate the most transactions, leading to the support costs distribution shown as Curve 2.

Curve 3 shows how total assets are apportioned in generating sales. It is easier to understand this curve if assets are divided into four pieces and considered separately.

First, inventories. The inventory turns on high-volume, limited-variety products are always faster than on low-volume, high-variety products. Hence, the inventory distribution to support product sales will be a shallower curve than the sales curve. While 80 percent of sales may be attributed to 20 percent of the product mix, the corresponding inventory numbers are probably split between 60 percent for the best selling products and 40 percent for the rest.

Second, machinery and equipment. High-volume products will require fewer changeovers than low-volume, specialized products. Specialized equipment may be needed for the low-volume products on the high end of the variety scale. So again, the machinery and equipment curve is probably a 60/40 curve.

Third, buildings and land. These are needed to store the inventory and keep the machinery dry. Hence, this curve follows the inventory and machinery curves.

Fourth, accounts receivable. For most companies, days outstanding is not affected by the type of product. Days outstanding is probably more a function of the customer.

Assume that the distribution of accounts receivable follows the same pattern as the sales curve (i.e., an 80/20 distribution). If different customers buy high-volume, low-variety products than buy low-volume, high-variety products, this assumption will need to be reconsidered.

These four factors—inventories, equipment, buildings, and receivables—yield a distribution of cumulative assets used across the product line, similar to Curve 3. This curve does not represent how the company had to invest in assets. When the first product was launched, there were probably some buildings and equipment that were not fully utilized. Hence, new products could be added without requiring an increase in these assets. The curve represents instead how assets are being used today to generate sales; one must be mindful of what the curve represents in interpreting it.

Curve 4 is calculated from the first three curves. For the first product on the variety axis, annual sales are taken from Curve 1. From this we subtract the unburdened direct labor and direct materials to make the product (assumed to be a curve of similar shape to the sales curve, that is, an 80/20 distribution) and, using Curve 2, subtract the support costs incurred by this product. The result is the true profit contribution from this product. This calculation is performed in turn for each item on the variety axis; the profits are then added together. Curve 4 is the result, showing that some products are net contributors subsidizing the losers. That some products subsidize others is expected; location of the winners and losers on the variety axis might be a surprise.

Curve 5 shows the cumulative return on assets (ROA) contribution from the products. This curve is calculated by dividing cumulative profits from Curve 4 by cumulative assets from Curve 3 then converting to a percentage. As indicated by the curve, a financially optimal product mix exists with respect to invested assets.

Traditional Cost Accounting Is Misleading

Cost accounting systems do not show these curves. Cost accounting places many support costs into fixed or period categories, not into product costs, even though these support costs are incurred to support specific product manufacture or sale. Other support costs are arbitrarily allocated to products based on some chosen variable (direct labor, for example, or machine hours, materials value purchased, etc.). These allocations do not track where support cost is truly incurred. Hence the terms "gross margin" or "contribution margin" are meaningless without knowing how much support cost is incurred by a product. Poor decisions can easily be made based on this information alone.

For example, a company making consumer electronics was told by their accounting system that their highest volume product was a loser.

They looked elsewhere to manufacture this product at lower cost and chose Taiwan. The production line was shipped and set up. They were told that the specialty products, situated on the right side of the variety axis, had large margins and as winners should be kept in the U.S.

The following year profits dropped dramatically. The highest-volume product ran almost untended. It had little or no support cost. But it was allocated support costs through a labor burden rate. As it was the highest volume product, it also had the most labor hours, so it was soaked with a large share of the support costs. On paper, this product was made to appear unprofitable.

The specialty products had very low labor content so they were allocated a small amount of support cost. Cost accounting showed a large gross margin. In fact, this company made exactly the wrong decision. They should have kept the high-volume product here and sent the specialty products to Taiwan. Support costs in Taiwan are low, so they could have reduced the profit deterioration from the specialty products.

The company mentioned earlier with such an enormous variety of products that it could not defend its market against niche competitors was also fooled by its cost accounting system. This company makes standard products and special products. Because gross margins were higher on special products, sales representatives were "incentivized" to sell more specialty products. This company was willing to manufacture any product. And that is what made its product portfolio so large.

Analysis of incurred costs based on transactional analysis showed a picture very different from cost accounting's. Industrial engineering, a factory cost allocated through a direct-labor burden rate, spent 82 percent of its time on specialty products, only 18 percent on standard. But standard products accounted for 80 percent of the labor hours in the factory, so they were allocated 80 percent of industrial engineering's expense.

Purchasing and receiving department costs were allocated to products based on a percentage of the material value purchased. Standard products use 80 percent of the material input, so they were allocated 80 percent of the purchasing and receiving budget. But analysis showed only 50 percent of the purchasing and receiving department costs were spent buying and receiving standard products. The balance was used for the specialties.

Order entry was a fixed or period expense. No attempt was made to relate this cost to products. Transactional analysis showed 95 percent of these costs were spent on special products, only 5 percent on the standard. The gross margin calculation made no allowance for this biased distribution of fixed costs.

The upshot of all this is that the company was "incentivizing" sales of the wrong products or severely undercharging for the specialties. They are now mulling over what to do based on this new information.

Eight Tough Choices

A quick, emotional reaction to the foregoing discussion could lead to only one conclusion: Chop the product line to maximize profits. Nothing would be further from the truth. The objective of the curves is simply to show that variety can be positive or negative. And some negative influence may even be positive—if the losing products stimulate sales of the winners. The loss leader idea works well in retail stores; it can also work well in other selling arenas.

The conclusion is not to automatically reduce the product line. Instead, we need to look at variety as a strategic issue that must be managed well or it will manage the operation.

These are the eight tough choices:

1. *Do nothing.* Either variety has not increased to the level where it is stealing profits, or losers are needed to gain winners. This is a very dangerous choice but is the most commonly chosen. It is dangerous because niche competitors can attack high-volume, low-variety products with significantly lower prices. The company following a do-nothing strategy cannot afford to reduce prices to match the competition because overall profits will drop or go negative.

Beware of the "full sales line" marketing mentality. Sales may feel they need a full line to serve customers. Call their bluff. Ask to see a list of customers who buy a complete range of products and how much. In other words, justify the loss of profits from some products to get the sales of others. Make sure this is a good return on investment before accepting the sales cliche.

2. *Price to suit total incurred costs.* The transactional concept of costs should show a better picture of total product costs. Price accordingly.

One of three things will now happen:

- Customers accept the change. What this does is alter the shape of the sales curve. Instead of an 80/20 picture, it now becomes 70/30 or even 60/40. The profit decline is less, or it drops to zero, or the company becomes profitable over the full product range.
- Customers reject increased costs on the low-volume, high-variety products but continue to buy the high-volume, low-variety products. The customers are now truncating the line. The next step is to attack the support costs and assets as discussed in choice 3.
- Customers reject increased costs on the low-volume, high-variety products, and because they *do* want a sole source of supply for the product line, they reject the high-volume, low-variety products also. Because of this risk, pricing needs to be adjusted carefully. Test customer reactions before going too far down this road, but also realize there are tremendous benefits. The value of recovering full costs goes directly to the bottom line.

3. *Truncate variety.* Deliberately prune the product line of obvious losers. Every company has "dogs" in its line, items that do not pull their weight. Phase them out.

The candidates for phase-out could be yesterday's winners that are still hanging on. Others might be products that were introduced with high expectations but never made it. Whatever the reason, they are probably incurring costs higher than revenues. Dump them. Any product-line pruning, unless it is a regular routine, should be done in phases. Customers, sales personnel, and others in the company need to get accustomed to the process. Launch Phase 1 with the obviously unprofitable items. A few months later, launch Phase 2 to reduce items that may be a little controversial. Leave the tough nuts until last. In other words, sneak up on them.

At this point the company needs a routine quarterly or semiannual process when everyone goes through the pruning exercise; otherwise the product line will slowly grow under the pressures of customers, sales people, and designers. Variety must be managed in both directions, increases and decreases. Reduction of variety will hurt sales revenues to some degree, but without a concerted attack on the support costs, profits will go down, not up.

There are two basic options. One is to reduce support costs as the transactional volume decreases. This is not easy because many department heads are unable to identify which activities should decrease as the product line reduces. The transactional analysis method applies here; otherwise Parkinson's Law will prevail.

The other option is to reallocate support costs to the remaining product lines and boost their volume. If this is possible, the effect will be to achieve the financial performance of the first of these two options without a reduction in support costs.

4. *Set minimum order and production batch sizes.* According to the transactional theory of costs, one way to reduce transactions is to set minimum ordering and manufacturing quantities. Because the penalty for minimum manufacturing quantities is extra inventory, the company needs to be sure before setting minimums that this will give enhanced ROA.

Many companies are sure that small customer orders are more costly to process than large ones. Setting minimums means the product is still offered in line but customers must want the item enough to accept higher inventories. Address the support cost issue as in choice 3 above.

5. *Buy unprofitable items and resell them.* Many manufacturers offer products in their line purchased outside the company. Another manufacturer, who perhaps is better organized to make these products profitably, makes them and simply attaches the correct company brand name to the product.

Using this method, the selling price must now cover only the purchase price plus the support costs of buying, stocking, and selling the product, plus any incremental profits needed to get adequate ROA from these products.

6. *Restructure the business—volume versus variety.* A high-volume, limited-variety business is quite different from a low-volume, high-variety business. Separate these two businesses. Manage them as two distinct entities with the objective of making profits from both. At least the separation will stop hidden subsidies that always occur when these two product types are manufactured and sold by one entity.

7. *Redesign—get application variety without product variety.* Multiple application products earn all the sales revenues without high support costs.

An example of managing variety, a company making medical electronics decided to export to Europe when the dollar dropped against European currencies. It needed to produce European voltage machines.

The engineers duplicated the full product line, the only difference being voltage. Costs began to increase faster than the increased export sales revenues. The culprit was increased product variety. A discussion with the design engineers determined that the products could be made voltage universal for a small incremental standard cost. After the change was introduced, cost escalation immediately stopped.

Other types of variety can be addressed through the design phase. Ingredients, vendors, and processes can all be reduced in the new-products design stage, with similar benefits to reducing end-product variety.

8. *Minimize Support Costs and Assets.* Many ideas from the just-in-time (JIT) movement result in eliminating support costs, especially in the factory. Less material handling, fewer inspectors and stores people, faster changeovers and better-quality products all reduce support costs. Lower inventories and smaller production areas reduce the asset base.

The same attack can be made on engineering and marketing support costs. A European manufacturer of consumer electronics once charted the volume of engineering changes over the life cycle of one of their products. They assumed that the number of changes was directly related to the level of engineering support. To their surprise, the level was almost constant. There was just as much activity during the dying phase of the life cycle as the beginning. The reason: They dedicated engineers to product lines. So engineers, of course, continued to refine the design without regard to sales volume.

Several Japanese manufacturers cut off all support to a product once it declines through a volume threshold. They prohibit *anyone*—design engineer, industrial engineer, or marketing person—from working on this item. The product must be a true cash cow to retain support. Talented and valuable resources are assigned to new products.

WHAT CHOICE TO MAKE?

Each company will have a different strategy to handle the variety issue. Some will specialize in high-volume, limited variety; others, in specialty products. There is no set answer for a business. The key is to start the discussion. Understand the implications of variety. Make the choice that best fits business objectives and the competitive environment.

LOGISTICALLY FRIENDLY DESIGNS

The preceding discussions imply that product design is a critical element to utopian logistics. The *P:D* ratio picture shows the importance of controlled product variety. The *P:D* ratio leads to the subject of standardization, as does the issue of variety. The importance of managed product variety and standardization highlight the need to focus on and control the design process.

The easiest way of showing the importance of design to logistics is through case studies. Cases bring out the problems clearly and show how design could have been improved with a little effort.

Consumer Electronics

A European company designed a new line of stereo equipment. Its performance was far superior to the competition, costs were about the same as the competition, and aesthetically it was excellent. The stereo magazines rated it the best in the world.

That's the good news. Now for the bad. The product design was logistically unfriendly (see Figure 7–4). The product range consisted of 56 end configurations. Some of these were color and language variations. Twenty-two different subassemblies, specifically printed circuit boards, could be assembled into these 56 varieties. Six bare boards could be assembled into the 22 subassemblies through the selection of some unique components.

The lead time (*P* time) in months to procure, produce, and distribute this product is shown on the left of the diagram. Point zero is when the end consumer buys the product. The distribution time from factory to consumer was an average of three months, while inventory sat on the shelves of various warehouses and distributors. Final assembly took 1 month, subassembly took 2 months; and the unique purchased components, available from only one source in the Far East, took 10 months. The bare boards took 1 month. Hence the total *P* time was 16 months.

The first step in the subassembly process was to add unique components. Hence the 6 common bare boards were made unique 6 months before customer purchase. The 22 subassemblies were made unique to the 56 end products, also in the first process step of final assembly. The forecast of color and language had to be made 4 months before the customer demand.

FIGURE 7–4
Time-Phased Product Structure

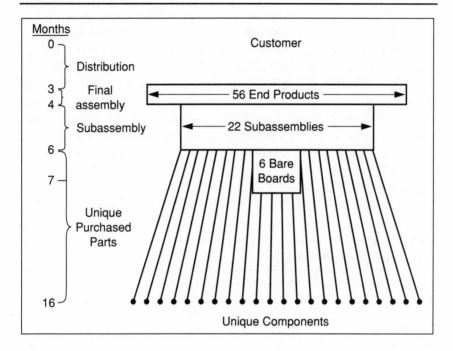

The result was a complex logistics problem. How could the manufacturer predict the right unique components 16 months before customer purchase? In fact, they could not. How could they predict which 22 subassemblies to produce 6 months ahead of the demand? Again, they could not. Worst of all, they could not predict which of the 56 end products to make 4 months ahead of sale. The end result of this design method was huge inventories of some end products and backorders of others. Backorders in this type of product means lost sales. Consumers switched to the second- or third-best available competitive product.

This company could have charged the competitors with marketing expense. The manufacturer stimulated demand for a product by their outstanding performance, but they could not fill the demand. The competitors reaped the benefits.

The number of projections needed plus the length of the forecast horizon guaranteed high unbalanced finished-product inventories.

The company tried frantically to get out of this condition. Expediting, air freight, overtime, products built without some parts that were added later, and idle time were all symptomatic.

A discussion with the designers showed that the product could have been designed to be logistically friendly as well as to meet performance. cost, and aesthetic objectives. Figure 7-5 shows the revised design.

The 56 end products now take on their unique identity at the last step of the assembly process. This trims one month off the forecast horizon. The 22 subassemblies are also made unique at the last process step, cutting two months off their prediction horizon. The unique components are selected from standard products, cutting their procurement time to two months.

The total P time from ordering these components to the consumer buying the end product is now 6 months compared to 16. Imagine what

FIGURE 7–5
A Logistically Friendly Design

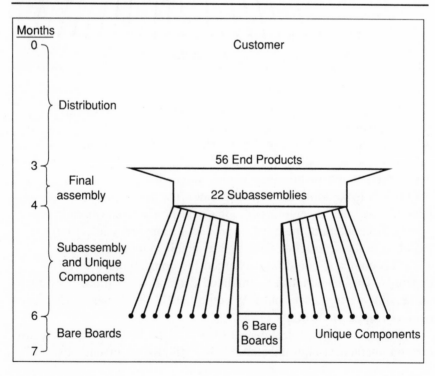

a logistically friendly design would have done to the success of this product in the marketplace.

This example emphasizes how the *P:D* ratio must be considered in product design. *P* time is especially important for products with short life cycles. In this case, consumer electronics life cycles are rarely as long as two years. A *P* time of 16 months over a life cycle of two years is a fundamentally invalid way of operating the business.

Six Key Objectives for Designers

Most design engineers believe they have three objectives to meet with any new or revised design. (1) It must perform a desired function (2) at a target cost, and (3) it must be aesthetically pleasing. Rarely do the objectives also include: (4) It must serve the customer with maximum product availability (5) with minimum inventories and (6) with maximum flexibility to market shifts.

The inherent design of a product controls 80 percent of a company's ability to perform the last three objectives. For most companies, these last three objectives are at least as important as the first three.

When the consumer electronics company's design engineers were challenged for having created a poor design, they answered that they had met their three objectives (Objectives 1 through 3). When asked about the second three objectives, they responded that they were not measured on these. In fact, they had not really considered them during the design phase. All new designs now meet the six design criteria.

Industrial Products

A midwestern company makes scanning technology used to monitor pulp and paper mills, aluminum foil production, cigarette machines, galvanizing lines, and the like. They build a variety of scanners to measure the critical characteristics involved in these processes. The devices collect readings and either display them for analysis and action or analyze them automatically and directly control the production process.

The company's problem is that none of the application markets is very large. Hence, the development costs to design products for each market must be spread over limited production runs. This means that either the costs per installation are prohibitively expensive or that profits are difficult.

Design engineers analyzed the product and sorted out what had to be unique to each market and what could be standardized. Their findings were that the scanners had to be unique but that the data collection and analysis electronics could be standardized. The standardized electronics would be more complex than any one application needed. With only one electronics unit for all applications, however, the total development costs would be lower. They also found that other support costs (e.g., planning, scheduling, purchasing, and accounting) would be lower, based on fewer transactions. They implemented the standard product strategy on the electronics business. The company is now trying to develop the same concept for the mechanical parts of the products that support the scanners.

Contrast this with a medical equipment company making diagnostic X-ray machines for hospitals. There are separate design teams for each product type. Even patient examination tables are different per product, as are the doctor's display units. Many of these and other items could be standardized between product types. This would, however, require assigning design teams to modules of products—tables and display units, for example—instead of to end products.

Because of the nonstandardized approach, profits are tight. Spare-parts support to such a vital, life-support product is very costly, flexibility to the marketplace is poor, and inventories are high.

These two companies are good examples of constrained versus unconstrained design activities. Remember that design engineering consumes only 5 percent of the total life-cycle costs of a product when the conceptual design is complete, but this 5 percent commits 85 percent of the total-life costs (see Figure 7–6). It is important to keep the 5 percent in control.

Product designs must be constrained to minimize total life-cycle costs. They must also be controlled to help achieve the logistics mission. Few designers are limited in this way. New product introductions need to be analyzed relative to this strategic issue.

Six Steps to a Design Architecture Statement

A design-architecture statement must address all logistical problems. It is a guide for design engineers. When confronted with design alternatives, it helps them pick the right one for the success of the product.

FIGURE 7–6
Product Life-Cycle Costs

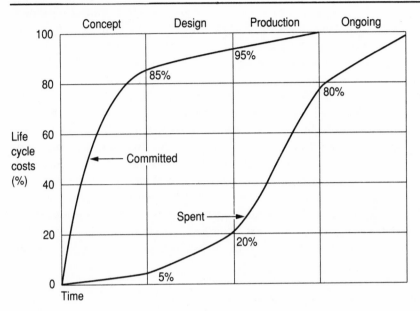

Step 1: Develop an Accurate Market Needs Statement
A product can have a wide or narrow market. The breadth of the market
and how much of it a product is designed to reach must be clearly
defined as a first step. For example, electronic products can be designed
for consumers or professionals for high, middle, or low price ranges
and can be suitable for the international market or for a few selected
countries. Industrial products can be targeted for all possible applications
or limited to certain specific industries. The scanner company clearly
defined which industries it considered its target market. This focus on
key market areas allowed the designers to create a single electronics
package to satisfy these markets.

Defining market needs is a difficult but necessary first step.
Looseness in the market-needs statement will generate an overly complex design with as many add-on features as the market asks for. The
proliferation of add-on features will be difficult if not impossible to keep
logistically friendly.

Step 2: Consider All Six Design Objectives
The six objectives defined for product designs need to be clearly understood by the design engineers and anyone else influencing design decisions. Each of the six objectives must receive equal weight; performance criteria need to be based on them.

Step 3: Develop Multiple Use Products
A broad product line is difficult to manage. One method to combat this is to design products with application variety without product variety. The medical electronics company selling in America and Europe, mentioned earlier, is a good example, as is the scanner company that took the same multiple-use approach with its electronic data collection and analysis unit. A slightly more complex unit than needed for any one application allowed the product to serve a wide variety of applications with a single design. Forecast accuracy, flexibility to demand shifts between markets, and inventories to support sales are all remarkably improved.

Remember Utopia: It is one product with infinite application capabilities. Move toward this objective.

Step 4: Control Component Selection
Components and raw materials must often be purchased or made in the early part of the P time for a product, considerably before the product's D time. Their selection is therefore critical. Classify and control them within the following limits:

- *Standard:* items readily available from a wide variety of sources.
- *Common:* items used across a wide variety of products.
- *Quickly available unique:* all items unique to a product having a very short lead time.

Step 5: Design for Short P Times
The lead times of procurement and production must be a design concern. They must be considered at the same time as the manufacturing process and the procurement alternatives. Short P times designed into the product will pay dividends to the company.

Step 6: Create Variability at the Last Possible Moment—Grow Mushrooms

Designs must be standard throughout the early part of the procurement and manufacturing process, with all the variability added as late as possible. A good example of this is dishwashers. Some companies include in the shipping box colored plastic sheets for all their offered colors. The consumer selects the desired color, inserts the sheet into the door and discards the rest. Contrast this to earlier designs in which color was a factory-controlled variable. Finished goods inventories were always out of balance due to the wrong color being in stock.

Figure 7–7 illustrates what is known as a "mushroom product." The addition of variety at the end of the process will eliminate many logistical problems.

Formalizing the Statement: The Design Review Meeting

The six points listed above are universal requirements for every manufacturer. The design-architecture statement will also contain items unique to the company's product or business strategy. The key point is to formalize this statement and hold regular design reviews to ensure compliance.

Some companies create a list of components approved for use.

FIGURE 7–7
The Mushroom Product

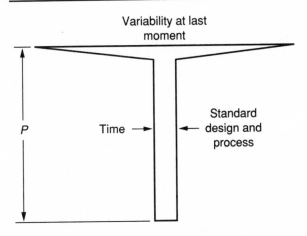

Deviations from this list are accepted, provided the designer can prove to his peers that the design will be compromised by staying within the list's confines. This adds a self-policing activity within the design group. It also ensures that the list is changed to keep up with new technology. The architecture statement will stop many of the futile justification exercises for standard cost increases.

The design review is the place to ensure logistically friendly designs. It should be conducted with representatives from sales, product planning, design engineering, manufacturing, accounting and purchasing. If the design process has been done correctly, most of these representatives will have been intimately involved with the design from its inception. They will have provided their insights into the product already.

The review formalizes agreement on the design among the key business team members. It ensures that the design-architecture statement has been followed. The critical item checked by the review is that the design will *not* fail because it ignored the key concepts of logistics.

PREDICTABLE SALES

The third strategic need from the *P:D* ratio discussion is to forecast more accurately: Limiting product variety and developing logistically friendly products helps the forecasting process. There are other factors imposed on the marketplace that already make accurate forecasting a difficult, if not impossible, task.

Our initial utopian manufacturer's mission statement idealized that raw materials arrive at the plant regularly and are processed immediately and that products are delivered directly to customers. Little, if any, inventory is in the logistics chain. To achieve this condition, demand for products must be reasonably smooth and repetitive. Huge demand swings make the utopian ideal difficult to achieve. Forecasting in a volatile marketplace is a challenge.

However, few sales departments consider smooth, regular demand as one of their objectives. Instead, many typical merchandising policies and programs encourage the opposite; they amplify any underlying market dynamics. The negative effects of these programs need to be understood and challenged. It's important for management to question the methodology used to stimulate demand.

The business will perform better if it can respond to true market-place dynamics without amplifying them.

Periodic Sales Targets

A large computer company based in the northeast has a regularly repeating sales pattern that destroys its ability to achieve a good logistics flow. In the first month of every quarter, it sells and therefore must deliver 20 percent of the quarter's revenue target. The second month, 30 percent. Fifty percent of the quarter's sales are shipped during the last month. A look at weekly sales within each month finds 10 percent of the month's sales are made in the first week, 15 percent in the second, 25 percent in the third, and 50 percent in the last week of the month.

Obviously, the demand patterns for large, expensive computers are not naturally this cyclical. This is an induced sales pattern caused by periodic revenue targets. Customers have been trained to order in this manner. They get better deals at the end of the month and at the end of the quarter. Customers are rewarded for ordering erratically with free software and maintenance arrangements and terms.

The factory subjected to this ordering pattern is forced to use inventories to decouple erratic demand from smooth factory operations. The buffer inventories are built early in the first month of the quarter according to a forecast of what customers will buy. Reconfiguration programs late in the month and the quarter produce the right products customers want out of the wrong inventory. Overtime is high, morale is low, vendors are jerked around, and product quality suffers. Billing errors are huge because of the scramble to issue end-of-period invoices. Errors are also caused because of the end-of-period deals salespeople make that grant many exceptions to their regular billing process.

Forecasting is obviously difficult in this environment. The wrong products get built, and the wrong materials and components are purchased. It is a wasteful, expensive process.

Poor Discount Structures

A midwestern company in the paper processing business used to offer steep purchase discounts to its customers. The lowest sales price was achieved if a customer purchased four hundred pallets of one material at a time. Only a few customers had enough volume to take advantage of this. The company could not predict when an order would arrive, and consequently never had stocked four hundred available pallets. Every unit was pulled from stock to try to satisfy the order, with the balance put on backorder. Crisis expediting in the plant, obviously at incremental cost, occurred as production attempted to make more of this product so it could be shipped at the lowest price. Orders from other customers for products on backorder could not be filled. All this happened even though the orders were entered at a higher sales price. Lost sales and customer dissatisfaction were common.

Artificially Induced Marketplace Dynamics

We have examined two very typical examples of a marketplace rendered dynamic by internal policies and programs. Other procedures requiring evaluation include billing practices, credit terms and conditions, advertising programs, and promotions. Management needs to monitor demand-stimulation procedures. The objective is to stimulate smooth, predictable demand.

One approach is to measure the sales ordering pattern and set objectives for the sales people that will smooth it. In the case of make-or assemble-to-order products, it is important that the request dates, not the promise dates, for a product produce a smooth demand. The request dates are the customers' need. If the request dates can be smoothed, it will be easier to deliver products when customers need them.

Smoothing demand at the same time it is stimulated will take ingenuity on the part of sales and marketing. One approach is to measure sales people daily instead of monthly or quarterly. Using different end dates for the measurement periods for different groups of sales people can also help smooth out demand. Rewarding a customer with sales discounts based on his annual order volume and how this is spread over the year pays off. This is a strategic issue that requires creative implementation approaches.

SUMMARY

The key business question is how logistics, the effective movement of raw materials from vendors through plants, can put quality products into the hands of customers. Companies that perform this job well have a competitive edge.

Logistics is a company-wide activity; all departmental functions have an impact. No department—staff or line—can be neutral. Just-In-Time (JIT) methods have accomplished much to improve product flow, concentrating largely on vendor relationships and factory activities. These techniques are not in themselves enough to give a competitive advantage. We need to broaden the scope of improvement to include product variety, design, and sales activities.

Companies that view logistics in a holistic manner make real strides. The synergistic effects of simultaneous changes in all areas can devastate the competition. Strategic decisions supported by strategic operating practices will really pay off.

BIBLIOGRAPHY

Mather, H. "Are You Logistically Effective." *Chief Executive*, November–December, 1987, pp. 32–36.

Mather, H. *Competitive Manufacturing*. Englewood Cliffs, NJ: Prentice-Hall, 1988.

Shingo, S. *Study of Toyota Production System*. Cambridge, MA: Productivity Press, Inc., 1988.

CHAPTER 8

HUMAN RESOURCE
MANAGEMENT STRATEGY

Robert E. McInturff

Editor's note

The 1990s have been called the personnel decade. Although the "mechan-ical" issues (JIT, TQC, etc.) have been well developed, the formulation and execution of an appropriate human resource strategy is often not seen as a necessary component. This chapter addresses what makes good human resource management (HRM) strategy, and takes us through three success stories.

Manufacturing strategy is ultimately measured by its successful implementation. Much has been written over the years about the impor-tance of human resources to the success of an organization, but not all organizations are as successful as others in implementing new programs or concepts and maintaining a competitive edge in today's world. Why do some succeed where others fail? And why is it that a successful management team from one company or division will move to another and not achieve the same level of success?

WHAT IS STRATEGIC HUMAN RESOURCE MANAGEMENT?

By the simplest of definitions, strategic human resource planning is having in place the right number of people at the right time with the right skills to meet company goals. Human resources can play an effective role in the planning and execution of corporate strategy through development of a human resource plan. The process requires a simple conversion of production requirements to labor-hours, followed by specification of further staffing requirements for training, development, and review.

HOW TO DEVELOP A STRATEGIC PLAN FOR HUMAN RESOURCE MANAGEMENT

Step 1. A Product Plan Converted to Personnel Requirements

It is a fairly simple task to convert a production plan or forecast to a personnel plan. The product requirements are converted to labor-hours using standards typically available in the cost accounting or materials requirements planning (MRP) systems. The hours are factored with efficiency data and converted to headcounts or labor hours.

Step 2. Short-Term Issues

Once the headcount numbers have been calculated, the employee and supervisory mix needs to be determined. The following questions should help complete the list of requirements, which are then compared to actual staff levels.

1. What is the right number of employees to meet business goals?
2. What job skills will be required to meet the goals?
3. Where are the people to come from?
4. What level of education will be required of the employees?
5. What level of ongoing training will be required to maintain the proficiency of the work force?

Complicating development of the strategic plan is that companies are constantly changing. Sales go up and down, new products are introduced or phased out, and employees come and go. At times the rate of change varies from high to low to barely perceptible, but change is perpetual. Although we can measure the *rate* of change by quantitative methods (i.e., sales dollars, number of employees, capital expenditures, etc.) the dynamics of change as it relates to human resources cannot be as easily quantified. But they are as real and measurable as any tangible asset, computer document, or numbers blessed by the accounting department.

THREE CASES OF EFFECTIVE IMPLEMENTATION

People versus Techniques

A manager of a 30 million dollar company states, "By cutting lead times, creating work cells, and improving vendor delivery programs we have reduced our manufacturing cycle times from months to days." This statement, as usual, focuses on the *techniques* that were successful but leaves out the *people issues* and the implementation details.

Our purpose is to identify characteristics that ensure effective implementation. We will look at three companies as examples of very different business environments operating successfully. Each of the three companies fits one of the following descriptions of internal change.

1. *Rapid or New Growth.* This condition is most often associated with a start-up. These companies experience rapid growth in both sales and number of employees. Their products are generally technologically advanced and require a high degree of sophistication in the manufacturing process.

2. *Loss of Market Share or Profitability Due to Competition.* Companies experience problems maintaining growth and profitability as their market share erodes. Their future growth is dependent upon implementation of new techniques, introduction of new products, and changing the management and performance measurements currently in place.

3. *Restructuring Due to a Merger or Acquisition.* Perhaps one of the most common occurrences in American industry today is the creation of a new business entity by a merger or acquisition. Companies are sold and merged, product lines are split off, and joint ventures are attempted in unprecedented numbers. The end effect is that while the plant and equipment, product, employees, and customers may all stay the same, the *corporate identity* changes overnight. The results include employee layoffs, partial plant closings, introduction of new products from other divisions or merged companies, and, most important, loss of the employee's vision of the company identity.

Below are the stories of three companies experiencing the stress of change, each coping in a different way. Their success stories are not unique, but as you will see, each followed the same road map to effective implementation.

CASE 1. RAPID OR NEW GROWTH—
THE START-UP: HI-FLYER ELECTRONICS

From its founding eight years ago Hi-Flyer Electronics has experienced a rapid increase in growth, achieving sales of $300 million with 1500 employees worldwide. The founder and president is a respected engineer well-liked by everyone. Although he is articulate, he is neither an impassioned speaker nor does he have a charismatic personality. People are comfortable in his presence. They feel they are listened to and their opinions matter. Profitability is excellent; sales are projected to continue growing at more than 25 percent per year. The manufacturing organization is well-run and has avoided the confusion and sloppiness associated with most start-ups. The vice president and all key senior and mid-level managers are still in place from the earliest days and plan to remain on board. Execution of the manufacturing plan is accomplished in a well-organized and timely manner.

Hi-Flyer's three major manufacturing problems are:

1. The organization needs to decide how to hire enough people to meet the continual staffing needs required to support rapid growth.
2. The company wants to ensure that old and new employees have the requisite skills to keep pace with the increased production schedules and the complexity of new technology introduction.
3. With no historical precedent controlling policy and procedures, how can the maintain a common goal throughout the organization? The number of new employees added yearly exceeds the total employee population hired to date, a factor that increases the difficulty of building an internally consistent set of goals.

CASE 2. LOSS OF MARKET SHARE—
THE GOOD PERFORMER UNDER
ATTACK: METAL BANGERS ARE
US MANUFACTURING

Erosion of market share due to the inability to meet demand for shorter lead times was creating a major problem for this manufacturer of custom machinery. At one time, Metal Bangers Are Us Manufacturing owned the lion's share of the marketplace. In recent years new technological advances and foreign competition jeopardized their number one position. Design Engineering responded by introducing a new generation of equipment. This restored their lead in design, but manufacturing still could not keep pace. Located in a remote suburban area, their ten-year-old facility did not have an abundance of skilled hourly and technical workers.

The challenge facing the president of Metal Bangers was to maintain market share. He knew they needed to improve their manufacturing capability. Options were limited, however. The current plant location could not be abandoned and the number of employees could not be increased due to lack of qualified workers. The president realized that any change would have to be accomplished using the current management team and factory work force. Although he could not see a way out, he took a time-honored approach: He reorganized and got lucky.

He created a new Plant Manager position responsible for both design and manufacturing engineering, materials, purchasing management, and shop operations. The slot was filled by the current Manufacturing Manager, an assertive and well-respected employee.

Results

The newly appointed Plant Manager brought all functional department heads, as well as significant individual contributors, together to create a task force for the purpose of improving manufacturing response time. The result of the task force work was to reduce internal lead times from months to days. Products are no longer built and assembled piecemeal; they are produced on an assembly line. Sales have increased by 50 percent, partly because of lead-time reductions, while employee headcount has remained unchanged.

How were these improvements accomplished? Through a number of simple and direct programs. Although no one knew what the final cycle time should be for internally produced product, it was agreed that it should be better.

- The project team became interested in the just-in-time (JIT) approach, and began to implement many of its techniques.
- Design and Manufacturing Engineering were combined so that all new products were introduced more smoothly.
- An aggressive training program was implemented and quality programs were put in place.

As the project progressed and many successes were achieved, the employees began to take more risks. The biggest risk was that, although they produced a custom product, typically a job-shop orientation, they wanted to build it on an assembly line. The result of changing to an assembly line is that average time from start to completion for a final assembly has been reduced by more than 50 percent.

CASE 3. THE LEVERAGED BUYOUT— A NEW IDENTITY: WHO ARE THESE GUYS MANUFACTURING

A division of a Fortune 500 company for years, Who are These Guys Manufacturing was purchased through a leveraged buyout. As their main product came to the end of its life cycle, the new owners faced a number of serious problems:

- Tight cash.
- The potential loss of 50 percent of their sales.
- An unmotivated work force.
- The plant location made it difficult to attract employees.

In addition, the company had chosen to revamp its traditional approach to the customer by adopting an aggressive new product-introduction campaign. This required a shift from low-tech/low-volume production to high-tech/high-volume production. They began an aggressive manufacturing program to increase efficiency, improve quality, and improve work force morale.

Results
Despite the loss of their major product line, sales are up over 20 percent from when the company went private. Employee efficiencies are up by over 15 percent. As measured by the Human Resource Department, job satisfaction and employee morale is at an all-time high. Soon after the buyout, a new president was named and a new Vice President of Manufacturing was appointed from within.

Who are These Guys Manufacturing integrated the three necessary components of Strategic Human Resource Management:

1. Commitment.
2. Training.
3. Measurement.

THE COMMON DENOMINATORS

Although the problems each of these companies faced were different and unique, there are some common stylistic and management approaches utilized by each.

The Leader or Organizational Communicator

In each of the three cases, one individual was clearly identified as the focal point, the acknowledged leader. He was viewed as the individual whose abilities were critical to the success of the organization and the people within it.

Although clear-cut authority was vested in one individual (generally the president), a dynamic functional manager could achieve the same level of success on a less than company-wide scale. Authority, charisma, and visibility do not necessarily imply comparable communication skills. There are some charismatic leaders whom employees respect but do not follow. There are authoritative leaders whom people follow but do not agree with, and there are visible leaders who are recognized, but not successful. We generally equate charisma, authority, and visibility with success, but the hidden fact is that truly successful leaders are *organizational communicators*. An organizational communicator can be either charismatic or authoritative or visible, but they all share the following traits and business practices:

- The communicator has a written business plan that is given to every top- and middle-management individual. Everyone knows where the company is headed.
- The business plan is updated on a regular basis, and the changes are communicated to the entire organization.
- Although the communicator is usually the president, another individual could assume the role if he or she had enough visibility.
- "Turf wars" are lessened because all decisions are based on the need for everyone to achieve the business plan.
- Top- and mid-level managers have more autonomy in their decision making because the need to run ideas and problems up through the organization is reduced. People at lower levels are able to make better-informed decisions because they have a clearer understanding of their effects on the business plan.

• There are few hidden agendas within an organizational communicator company. If the company needs to develop three new products a year to be successful, it is common knowledge. If those three new products must fill a specific market niche, every one knows that also. Specific engineering specifications and detailed marketing plans may have been on a more traditional need-to-know basis, but an overall lack of secrecy permeates the entire company.

DEVELOPING A STRATEGIC HUMAN RESOURCE PLAN

Earlier we described strategic human resource management as having the right number of people at the right time with the right skills to meet company goals. How did the three companies institute such a plan? There appear to be two critical steps in the actual implementation of a workable plan. The first is *detailed planning at various functional levels*, and the second is an *intensive training plan supported by a budget*.

Companies with organization communicators are able to create a management plan throughout the entire organization. The plan starts from the top and works its way down throughout the organization, creating more detail at each level. For example, the Vice President of Manufacturing is aware that new product sales are predicted to grow by 50 percent. This involves the introduction of a new technology, so obviously it will have a major impact on the organization. The impact is reviewed in detail with the Manufacturing Manager, looking at its effect on manpower, capitalization, and potential organizational changes. The Manufacturing Manager then fleshes out the plan in further detail (perhaps with manufacturing engineering, purchasing, production, etc.) and provides even greater detail as to what will happen. Finally, the operating personnel break the plan down further and develop a detailed plan at their level.

The process itself is nothing remarkable; it is probably the same process undergone by almost every company. What makes this planning process different and unique is that *at every level*, everyone knows the company strategy—the whys and hows of what is being done. Each step of the decision-making process contributes to the health and vigor of the whole organization.

Prerequisites for Developing a Workable Plan

There are two overall conditions that impact the decision-making processes within these three companies: the ability to make autonomous decisions and the organization's support structure. The manager's ability to make his own decisions is dependent on the corporate plan, the source of credibility for this level of decision making. Within the organization there is a commitment to support people so as not to let them fail. This commitment to the success of the organization and the people within it provides checks and balances; no single plan or manager ends up off the mark. There is constant communication and consensus on direction.

The importance of the philosophy of commitment and consensus must not be understated. The people within the organization work hard on their plans and goals because they feel a sense of ownership. The effective organizational communicator creates an environment where management's actions are recognized and their plans are implemented. It fosters a feeling that what management does counts. It is a cycle: Management feels empowered and strives to reach corporate objectives; corporate objectives are reached, and that further reinforces or validates management. This reciprocity was extremely important in all three cases.

Checks and balances. Each department is allowed and encouraged to be in control of its decisions. But because all groups filter their decisions back through the entire organization, no one can get too far off track. The organization's members help each other succeed.

At Hi-Flyer Electronics, a department manager who is levels removed from the Vice President of Manufacturing could not conceive of being excluded from the planning process or of not making his own decisions on how to run the department. The corporate culture of autonomous decision making was so ingrained that the department manager could not even "imagine why I would work there or why the company would want me, unless I had a decision-making role."

DECISION MAKING

The attitudes that pervade such an autonomous decision-making environment are:

1. Managing by example. As a result of the organizational communicator's constant flow of information, employees feel it's only natural to respond with information of their own.
2. Managers are proactive. They identify problems, seek input, recommend solutions, and build consensus instead of waiting to be told what to do and how to do it.
3. Each employee is chosen because of his quality as a person. In other words, it is not a simple choice of skill sets. Each team member must be able to contribute tangibly (e.g., technically, financially), and intangibly (e.g., creatively, proactively, communicatively). Managers are hired in part because they expect to be autonomous decision makers.

Training

The biggest surprise among the factors contributing to the success of each of these organizations is the tremendous emphasis on training and the budgeting process associated with it. In talking about the organizational communicator and the planning process, an observer might wonder, "These are real people in real companies. How are they making this work? Why can't I do it too?"

The answer is that these companies used the training-budget process as the mechanism to drive human resource planning. Each company had to make sure its work force could keep up with new demands:

- Hi-Flyer Electronics needed enough skilled workers.
- Because Metal Bangers could not bring in new people, they had to make sure their current employee base was ready.
- At Who are These Guys Manufacturing a massive retraining program was implemented as they phased out one product and introduced many others.

The Training Budget

The training budget was key to how these companies were able to implement the strategic human resource plan. It is directly linked to the organizational communicator's strategic plan.

With an organizational communicator, short-term and long-term goals and objectives for the entire organization are known and understood. Planning is done at each level. Each department manager has a crystal clear idea of where the company and its departments are going and what this means for his department. The department manager has stated objectives that he must support with an action plan.

The Training Plan

There are six questions that must be answered.

1. What is needed to implement the plan?

First, the manager identifies what skill sets are required to make the plan happen. This forces him to review his current employees. Can he use the people in place through training? Could the required people be hired from the outside?

At Who are These Guys Manufacturing, a decision to implement statistical process control (SPC) was made. This required specific job skills. Although the company had managers who were certified to teach SPC, now they needed employees to practice SPC.

2. How many employees are needed?
3. Who will train the employees?
4. What do the employees need to learn?
5. When should the program be given?
6. Where should the program be given?

Who are These Guys decided the training should take place on-site during work hours for the following reasons:

- Running the program on site ensures that it gets done.
- The company is in control, not the teacher.
- This learning program becomes mandatory, not voluntary.
- The program becomes a visible statement of the corporation's commitment to SPC.
- Employees are more committed to the learning process.

Although providing this training during work hours causes inconvenience, the company believed that to ensure success of the SPC program, it had to be done during the work day. This training program was planned to be ongoing since the SPC program would stay.

Succession Planning

Succession planning is a well-known tactic in Corporate America, and most corporations make provisions for it to some degree. The larger the company, the more sophisticated the planning. However, what makes the organizational communicator's company unique is that succession planning is a taken-for-granted, one-more-thing-to-do part of everyday responsibility. Just as one must meet schedules or prepare budgets or give a performance review, so one must constantly prepare a succession strategy. The acceptance of responsibility for and the ongoing attention to succession planning at all levels of management was evident in Metal Bangers Are Us, Hi-Flyer Electronics, and Who are Those Guys Manufacturing. Continuous succession planning results in the right people being available at just the right time to satisfy corporate objectives.

At Metal Bangers the line supervisor planned to hire two skilled machinists in two and one-half years. Although a supervisor at this level would not generally participate in long-range planning, in a company with an organizational communicator, each professional manager could translate long-range objectives into their impact on his department. To meet the need, he chose to look internally for the talent. After planning and identifying two potential candidates for promotion, he also created two potential vacancies requiring replacements.

Succession planning is an integral part of this line supervisor's responsibility. Imagine walking up to the typical line supervisor and saying, "Create a succession plan." The response might be a baffled look. However, if you say, "In two years we will need two more machinists that we can't hire from the outside. Go figure out who you want to train," then you have asked for a workable succession plan.

MEASUREMENT

Finally, each organization set up performance measurements to ensure completion of the plan. The standards were not arbitrary and they were carefully followed. They included:

- "Hard" data rating actual performance against the plan on a corporate, departmental, and individual level. Actual performance against objectives specified in the plan measures projected benefits of, for example, using funds set aside for training and development of new managers, or releasing full kits from the stockroom to production.
- "Soft" data concerning employees' feelings on a corporate, departmental, and individual level.

Many people become suspicious and fearful when their work is measured. But the employees of the three "success story" companies welcomed the opportunity because they had participated in setting the standards by which they were measured.

People became excited about setting objectives, achieving goals, and being recognized for their accomplishments. The process was contagious. At the next planning session, expectations were raised as goals were set at higher levels and the company's success continued.

At Who are These Guys Manufacturing, the measurement process is seen as an instrument for positive recognition. Because of the lack of historical data, new-product introductions were used as an opportunity to introduce performance measurements. Working with a clean slate, management set a goal; once it had been achieved, they set a higher goal. Finally, when one of the hourly workers asked, "Why only count new products, why not count all products?" performance measures were born. Good ideas came from all levels.

SUMMARY

Although most organizations will not achieve the results of the three companies profiled, partial implementation is possible. So what can we do? The fundamentals discussed in this chapter—commitment, training, and measurement—can be broken down to a much smaller scale, either for individuals or departments. Although there may not be an organizational communicator within your company, that does not mean that a manager cannot be one within his own area of control.

What is needed is a strategic human resource plan for yourself or your department covering staffing levels, training and development, succession, budgets, and measurement. Let people know what it is. Create the opportunity for feedback through the plan to the organization. Determine organizational needs through training, and measure the results.

174

CHAPTER 9

DIGITAL EQUIPMENT CORPORATION: JOURNEYING TO MANUFACTURING EXCELLENCE

Patricia E. Moody

In 1983 Digital Equipment Corporation experienced "the big bang." A few shipments were missed, customer service slipped, stock prices fell over 46 points, and industry analysts claimed that the company was undergoing extreme growing pains. Internally, many long-time employees were shocked.

For years Digital had ridden a high-speed train of success driven by product innovation. Manufacturing and cost-control methods were less sophisticated than those of other industries; likened to the shoemaker's children who go barefoot, the company had been described by the old quote, "We don't use computers, we make them." Nevertheless, the company had successfully executed key strategies of getting technologically innovative products to market and establishing manufacturing and distribution networks to fill the pipelines.

Digital's response was to look at the problems and to focus on transformation of manufacturing into a powerful, confident partner in the product sales and delivery cycle. This transformation is the subject of this chapter.

More "traditional" companies have encountered the same crises. While they may have taken a chainsaw to the problems, cutting out large pieces of the organization along the way, Digital has managed to

TABLE 9–1
Burlington Plant Performance Results

	1985	1986	1987	1988	1989
Manufacturing cycle time (power supplies)	10 weeks	2 weeks	5 days	3 days	2.5 days
Master schedule performance weekly	75%	85%	97%	97%	daily
WIP inventory turns	<20	<20	23	49	65

transform itself without killing anyone. The process and the speed in which major changes were accomplished are impressive (see Table 9–1).

What is it that enabled Digital to accomplish massive culture change and emerge healthier and stronger than before? In this chapter we will examine the mechanical and people aspects of the DEC experience that highlight their innovative approach to culture change. Visible corporate commitment; organizational restructuring, training, and education; some reshuffling of the players; new manufacturing methods; and a new type of manager were all part of the process.

We will look more closely at one of Digital's manufacturing facilities, the plant in Burlington, Vermont, which manufactures power supplies and systems products. This plant is one of 35; it is not atypical of the company's new approach to manufacturing. Similar success stories have come out of other plants (e.g., Augusta, Maine; Albuquerque, New Mexico; and Enfield, Connecticut).

BACKGROUND

DEC Burlington is staffed to run full tilt at 1000 employees; it is one of the company's two high-end systems plants in the United States. The introduction and manufacturing of a new computer product is a complex process that requires coordination of many functions—engineering, software, marketing, and manufacturing. The cycle of new-product development and old-product decline causes significant swings between peaks

FIGURE 9–1
Digital Equipment Corporation, Burlington Plant: The Journey

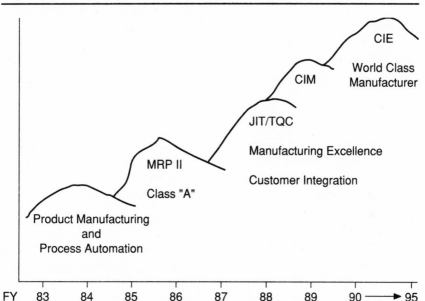

of market demand and valleys when the plant may carry excess inventory, staff, and capacity.

After the problems of 1983, manufacturing at Digital was challenged to eliminate its firefighting approach to production in favor of more planning supported by quantitative measures to reflect good and bad performance. For the Burlington facility, as well as for the others, the journey had begun (see Figure 9–1).

MANUFACTURING STRATEGY

Goal Alignment

The crisis caused a reevaluation of the alignment of manufacturing, corporate, and plant goals. Much work was done using a common unit of measure, metrics that moved beyond measurement of revenue ship numbers to data reflecting productivity in material management, scheduling, and shop floor control. Management felt the company needed to concentrate on doing the basics well, as reflected in the metrics. One of the first steps, therefore, was to seek Class A MRPII certification.

The Burlington plant themes, which are posted throughout the plant, include simplicity, elimination of waste, flexibility, cleanliness, continuous improvement, collaboration, and diversity. To reach a Class A level of manufacturing competence, seven areas had to be improved, each area containing specific goals for improvement:

- *Strategic planning*, including (1) production planning accomplished monthly by senior management from Engineering, Finance, Manufacturing, Materials, and Sales and (2) product-family production plans to support the master schedule function.
- *Operations planning.*
- *Operations control*, including (1) cycle counting to replace the annual physical inventory, and (2) a Class A appearance.
- *Data management.*
- *Performance measurements.* Thirteen were in place after three months, and the plant was working toward 90 percent-plus performance in each of the 13 areas, of which these represent five:

  1. Return on Assets (ROA): 90 percent of planned ROA.
  2. Sales: 90 percent of sales forecast.
  3. Production: 95 percent of production plan.
  4. Units Completed: 95 percent of units scheduled.
  5. Schedules released on time: 95 percent of schedules released.

- *Documentation.*
- *Education.* Company support of training and education programs was reflected in budgets to run them.[1]

In 1986 the Burlington plant was certified Class A, along with several other facilities. The extremely small number of certified Class A MRPII sites in the world (approximately 250, of which 10 percent belong to Digital) indicates the enormity of this task.

What Digital learned about culture change in the certification project they continued to apply as manufacturing moved beyond Class A MRPII into just-in-time (JIT), total quality control (TQC) techniques and philosophies. The key was not to work harder with more discipline, or to take the chainsaw approach, but to change the culture of the plant and to change the way the business was managed.

One key strategic decision, to keep the entire life cycle of the product in one plant, has major implications for how the facility deals with flexing inventory, as well as the work force. Smoothing the schedule, as well as inventory and manpower, is more difficult when product demand shows the swings typical of new-product life cycles. The organization and material management guidelines are intended to support new-product introduction, ongoing volume production, and product decline.

There are other innovative approaches to manufacturing excellence apparent at Digital. Let's look at the mechanical and the people aspects separately.

In the section entitled "Mechanics" we include production planning and control systems, quality control, and physical plant and equipment factors. What makes Digital's new approach to manufacturing exciting is not, however, its emphasis on capital expenditures and expensive computer software. It is the sociotechnical aspect of culture change, or the organization and people, that make Digital's work effective.

THE MECHANICS

The Plant

The Burlington facility is a new and pleasant environment to work in. The floor layout is clean and easily traversed. Work centers are set up as autonomous work cells (responsibility centers); many are fed by conveyor systems. The layout based on work cells, scheduling performed on a "pull" rather than a "push" basis, and reduced in-process inventories reflect this transition.

One of Digital's manufacturing strategies, process simplification, has allowed a bit of "reverse technology." A complex and expensive automated material-handling system containing some artificial intelligence capabilities has been partially dismantled.

Digital plants have moved beyond simple Class A certification to JIT/TQC philosophies and techniques. Focus has shifted from machine utilization and efficiency measurement to quality and leadtime—both external, customer issues.

The Planning System

Material requirements planning and some capacity planning is performed using a commercial package. The system is used to do parts explosions and determine vendor requirements; it does not use the modules for all the other order-launching and tracking activities typically offered in material requirements planning (MRP) packages. Before the 1983 crisis, forecasting and scheduling were what one manager called "an informal mess" complicated by numerous market/product groups, all competing within manufacturing for capacity and material. When the company reorganized, individual market/product groups were eliminated, enabling manufacturing to follow total corporate objectives. The goal was to use one sales forceast and one ship forecast for the entire company. Manufacturing plants therefore became suppliers to all markets. The architecture of the VAX product in essence dictated the new, unfragmented organization structure.

Quality

One of the requirements for successful JIT operations is excellent quality. Without the through-process yields achieved by TQC, pull-system lines are shut down; short-cycle techniques do not work. At Digital, major training efforts resulted in general understanding and use of statistical quality-control techniques by production personnel.

Quality-control inspectors have been eliminated; the focus of the quality assurance organization has shifted to quality at the customer site. Fifty percent of components have no incoming inspection. As the plant compiles more and better data on suppliers and builds closer relationships with them, parts performance and conformance flaws are reduced by separating supplier from design problems.

THE PEOPLE

Corporate Culture — Managing Change

Russ Snyder, the Burlington Customer Integration Manager, looks at their work on the dynamics of culture change this way:

This story is about managing change. It is not about hardware, software or automation, but people issues. This is the hard stuff. We work on the mechanics, but they are just a tool, we use them as a catalyst. To change culture is the final objective — to change the way the business is managed. We are positioning ourselves to continuously improve; that is what will ensure our future in the world class manufacturing arena. We'll be in the game.

The Digital work culture has gone through several generational adjustments. The new Digital manager represents a sharp contrast to the kind of start-up manager who successfully brought a mix of political skills, clout, and enthusiasm for firefighting to the company's first 20 years. The ideal new manager has evolved from a good soldier to a smart, creative professional.

Performance measurement. The emphasis has shifted to "metrics," valuing diversity and innovative approaches to people-empowerment (even the word "motivation" seems inappropriate here, as so much responsibility has been shifted to employees). Changing the way people are measured is the best way to begin changing their behavior.

The learning organization. The plant's working atmosphere reflects directly on the personality of its top manager. At Burlington, Jon Wettstein combines quantitative approaches with an optimist's belief in the creative strength of empowered employees. People are encouraged to go off on tangents and to take risks; they are rewarded with high visibility for good ideas.

As reflected in the contents of managers' bookshelves (including books by Tom Peters, Richard Schonberger, Shigeo Shingo, and Bob Hall) reading is encouraged. In fact, during the the JIT/TQC transition, groups of employees met in weekly study sessions to discuss and dissect texts chapter by chapter.

Digital has participated in a number of study missions to Japan. A movie theater, "The Burlington Cinemax," was set up in the cafeteria, complete with popcorn machine, to accommodate the Excellence Series, month-long explorations of new manufacturing ideas presented through videos, films, and live speakers.

JIT/TQC and the creation of work cells require solid team-building skills. There is much training in the "soft skills." The VIA (Vermonters in Action) program, for example, taught all levels of production personnel the details of statistical quality control (SQC); along the way the process was also structured to address problem solving and teamwork.

Managers and supervisors are now being trained in experiential programs like the Ropes Course, an Outward Bound-type session, which
takes place off-site. These programs are regarded as important in developing strong internal bonds to improve organizational effectiveness, not
just as perks or performance rewards.

Valuing Diversity

The plant staff is diverse—50 percent are female (as are 60 percent of
the production workers). Within the work cells there have been efforts
to reinforce the fact that diversity among employees means respecting
individual strengths. Some members of a cell, for example, may not
want to become skilled production leaders but may be most comfortable
and most effective performing specific technical jobs. Valuing differences requires a shift from individual "win-lose" approaches, to group
collaborative approaches.

Co-location

The new-product introduction cycle, which has typically taken years, is
being compressed. Co-locating key design and manufacturing personnel,
either at the plant or at design centers located in Massachusetts and California, helps to speed the process and reduce postintroduction problems
like engineering change orders.

Other design-for-manufacturability issues receive attention as a
result of this collaboration. Tolerances for integrated circuits, the
specifications of which are used by purchasing to set up vendors,
are realistically chosen to reflect quality as well as manufacturability objectives. For example, tolerances that are tighter than product-
performance requires and that consequently induce higher failure rates
are examined closely to be sure that they are absolutely necessary.

Outward Focus

Customers and other VIPs are a frequent sight on the production floor.
There is constant exposure to customer presence, which according to
Snyder ". . . reminds us of why we are here." Customers who visit the
plant take part in peer-sharing sessions in which Digital personnel review
their manufacturing improvement programs, discussing failures as well
as successes.

WORLD CLASS MANUFACTURING

Employees are aware of the fact that approximately 50 percent of revenues come from outside the United States. The next objective on Digital's list of achievements is to reach world-class manufacturing status (see Figure 9–2). Although the exact definition of "world-class," is still being formulated, some of the desired characteristics have been identified.

According to Doris Adams, a consultant working with plant personnel to prepare for the next phase, the following four characteristics describe world-class leadership:

- A knowledge base of what it takes to be world-class.
- A sense of the business—horizontal, vertical, and global (e.g., go abroad and bring customers in).
- An ability to work with people collaboratively.
- The ability to frame a vision and hold it up for testing.

SUMMARY

Over the years, Digital has demonstrated tremendous skills in change management. When facing similar crises, other manufacturers in high-tech industries (e.g., Data General, Honeywell, Wang, RCA, Osborne, GE) have not done as well.

Digital and other high-tech companies face strong competitive challenges in the next five to ten years. High-tech industry is maturing. It is clear that competitors, particularly those in the Far East, have targeted the computer market.

Digital's toughest challenges now are (1) the risk of getting stuck with excess capacity (e.g., manufacturing personnel) and excess inventory and (2) maintaining a leadership position in new technologies. In future interconnect technology areas, for example, competitive manufacturing capabilities in surface-mount and tab technologies for printed circuit boards will have to be developed.

So far the company has been able to adhere to its "no layoff" tradition. As product technology and the markets have shifted, large numbers of employees have been relocated or reskilled.

The treatment of inventory as the "last cushion" against supplier or forecast problems is being addressed. Traditionally, new-product

FIGURE 9–2
Digital Equipment Corporation's Time-Phased Journey to Manufacturing Excellence

Goal	Phase I: Basics with Excellence in Manufacturing	Phase II: Standard-Setter in Computer Industry	Phase III: Leader in Manufacturing; World-Class Competitor	Phase IV
Strategies	MRP II	JIT/TQC	CIM	CIE
Shortest Cycle Time in the Industry	MPS execution Vendor scheduling Closed loop system Simplified processes	Queue time Lower setup time Continuous flow Cut flow distance Eliminate space Eliminate repairs Reduce lot size Quality at the source Order administration system	Networking Super MRP II (10-minute MRP simulation) Continuous flow Artificial intelligence Transaction elimination DRP II	Expert system configuration aid OTP Artificial Intelligence Ship direct to customer Customer on-line order administration
Competitiveness Independent of Volume/Time	Zero defects to customer's standard 24-hour shipment to customer	Zero defects Customer gets shipment when needed/wanted Remove buffer inventories Lot size of 1 Make item every day	Single-Level BOMS Automated material handling Bar code Optical readers Clustered computers Accurate data Manage capacity Pull software	Knowledge network Administrative systems Material architecture Automation Data highways Data resource management

25% of Manufacturing Market	Leverage class "A" Class "A" MAXCIM Customer reference site Manufacturing corporate account managers	Know your customer Flexibility Manufacturing reference sites Manufacturing corporate account managers	Know your customer Customer on-line to DEC systems Plant demos Customer consulting	Time to market
People	Educated/trained Productivity measurement Stride	Multi-skilled Employee involvement Problem-solving techniques Stride Job plans Career changing	20% time always in education Manufacturing teams Stride Education sabbaticals	Multifunctional teams Stride
Products/Process That Never Fail	Meet delivery commits Uninterrupted power Housekeeping Redundant systems SPC	Remove repair loops Perfect incoming quality Continuous flow Quality at the source	Fault-tolerant computing Data resource management	Minimize steps (handoffs) from design through ship and aftermarket service Data resource management
Method of Manufacturing	Discrete batch	Flow		Flexible
Time Reference	Weekly	Daily		Per shift or less (hourly)

introductions require heavy inventory investments. But with turns levels running at 4, targeted for 9 or 10 in the 1990s, Digital is moving in the right direction.

The nature of the business has shifted from a focus on profits derived from hardware sales, to other related product offerings—software, communications, and field service. As equipment margins get tighter, manufacturing needs to decrease overhead costs. The trend is for fewer direct-labor and middle-management personnel in production.

In 1983 Digital looked at a major performance problem and used it as the catalyst to mature its systems and change its manufacturing strategy. The issues we have examined, the tactics that support that strategy—introduction of performance metrics, the evolution of a new management ideal, the empowerment of employees, innovation in the "mechanical" aspects of the business (MRP, SQC, etc.)—are important indications of the company's ability to survive, improve, and compete.

The decade of the 1990s has been labeled the personnel decade. Certainly Digital's creation of work cells and its atmosphere of training and learning, are appropriate for that era. The second major challenge, the size and make-up of the workforce, will hopefully be addressed as sensitively and creatively as have other people issues. If it is true that the values of the president color the values of the culture, then we can be optimistic about the resolution of this problem. President Ken Olsen's record of concern for people, his humanistic approach toward growing the company, will hopefully lead management to the right decisions.

What can American industry learn from the Digital experience? There are good and bad points to consider.

- It took a crisis to get management's attention.
- Technology industries need to work on compression of new-production introduction cycles.
- Total Quality Control (TQC) at the production worker level is achievable.
- Creation of the learning organization is exciting and important; strong competitors in the next decade will be turning their attention to better utilization of human resources.

NOTES

1. "Class A Business Resource Management Check List" (Hunt Valley, MD: Alban Associates, Inc., 1986).

PART 4

TACTICS

Part Four, "Tactics," contains specific examples of tactics that are applied to support manufacturing strategies. The new emphasis on speed, getting to the customer fastest, is an exciting challenge explored by Joseph Blackburn. Julie Heard's piece on JIT (just-in-time) for white collar workers also addresses the issues of how to compress process time at the front end of the cycle. The Goulds Pumps case illustrates her points on white collar productivity. The two chapters by Mark Smith and Richard Bombardieri look at case studies in three different industry environments: mining machinery manufacturing, electronics, and pharmaceuticals. Each of these chapters contains specific recommendations for productivity improvements. Finally, in Chapter 14, Charles Fine of MIT outlines important developments in new manufacturing technologies.

CHAPTER 10

TIME-BASED COMPETITION

Joseph D. Blackburn

Editor's note

Time-based competition is a logical approach to the next frontier in competitiveness. Time compression, looking for improved customer service through quick response, results in manufacturer cost savings. This chapter explores the whole exciting topic of time-based competition, lists the successful innovators, and will help your company take non–value-added time out of your delivery cycle.

In their struggle to regain a competitive advantage, U.S. manufacturers have too often neglected to use their most potent weapon: time. Physically located within the world's most lucrative market, U.S. manufacturers should have an overwhelming advantage in terms of responsiveness to customers. Unfortunately, rather than exploit this advantage, many U.S. firms are in the uncomfortable but familiar position of observing Japanese firms demonstrate the significance of time as a dimension of competition. In the process, these Japanese firms are successfully negating the geographical advantage held by U.S. firms.

TIME AS A COMMODITY AND COMPETITIVE WEAPON

According to Fred Smith, the Founder and CEO of Federal Express,

Time—as both a commodity and a competitive weapon—is an emerging issue that business people can't ignore if they expect to survive in this increasingly competitive world. We will see the demise of marginal firms who do not adopt time-based strategies. And the longer they wait, the faster they will fall. In short, where everything else is equal, time-based strategies become a key factor in widening the gap between those who adopt them and those who do not.[1]

Time is indeed displacing cost as a critical dimension of competition (see Figure 10-1), and firms who ignore the importance of time in their own processes are clearly at risk. The potential vulnerability of traditional manufacturers is clearly illustrated by the following statistic. In the average firm, *less than 5 percent* of the total time required to manufacture and deliver a product to a customer is spent on actual work. The remaining 95 percent is non–value-added time and this wasted time represents a gaping window of opportunity for the time-based competitor.

The just-in-time (JIT) revolution of the past few years was perhaps the first manifestation of a new type of time-based competition. Yet JIT is not an end in itself for a firm but an evolutionary step toward the long-term goal of total time compression. Zero inventory and proximity to suppliers, while important, have only a limited impact on the bottom line. The real benefits—the ones that provide a sustainable competitive advantage—are created by a shrinkage of the whole manufacturing cycle. This type of time compression translates into faster asset turnover, increased output and flexibility, and satisfied customers. Viewed in this way, the diminished inventory often associated with JIT is seen as more of a side benefit than a driving force.

FIGURE 10–1
Progression of Competitive Advantage in Manufacturing

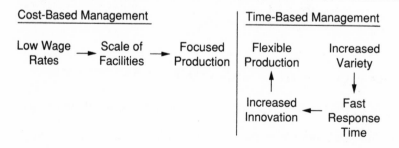

What is Time-Based Competition?

The phrase "time-based competition" originated with George Stalk and his colleagues at the Boston Consulting Group. In the research for their book, *Kaisha: The Japanese Corporation*, Stalk and Abegglen observed the evolution of JIT at companies such as Toyota. In their view, time-based competition is the extension of JIT principles into every facet of the product delivery cycle, from research and development through marketing and distribution. When one firm achieves such a significant advantage in product or service delivery, the nature of competition in its entire industry is changed: Cost becomes secondary to response time. Companies in the textile and apparel business have dubbed this new capability "quick response."

IDLE TIME EQUALS WASTE

Both JIT and time-based competition have identical objectives: Eliminate all waste in the production of a product or delivery of a service. With JIT, small production runs, quick changeovers, and low inventories enhance production speed. The percentage of time when value is being added to the product is maximized. This is also the essence of time-based competition. Eliminate idle or dead time wherever it exists, make sure that work can be processed in small batches, and maximize the value-added time. Time-based competition, however, goes one step further, encompassing not just manufacturing but the complete product cycle.

Order Processing

Consider the order-processing function. In many cases it takes longer for the firm to process the order and place it on its master schedule than it does to manufacture the product. Most of the processing time is wasted time. No value is added while batches of paper fill in-boxes. The reasons for delay are frequently the same as in manufacturing: Information is processed in large batches; orders are moved in batches from one location to another; large inventories of orders await some form of approval; processing of orders is sequential rather than parallel.[2]

To compress order-processing time requires actions almost identical to the changes needed for JIT implementation in manufacturing:

- Simplification of the process itself, cutting out all unnecessary steps.
- Better tracking to reduce "dead spots."
- Cross-training of workers to eliminate single-minded focus on a small part of the total process.
- Higher in-process quality to eliminate checking and rechecking (inspections).
- Group training.
- Running some efforts in parallel rather than sequentially (e.g., Simultaneous Engineering).

The implication, then, is that firms who are or have been successful with JIT will have a leg up on their competitors in time-compressing other phases of the business. This notion, however, may be wishful thinking. Many companies have not started to look at the front end of their process. Of those who have, the area is still relatively new and filled with opportunity to try out JIT manufacturing ideas or to develop new ones to speed the process.

Northern Telecom, a leading manufacturer of telecommunications equipment, found that although it had dramatically compressed time in one manufacturing process for large digital switches, months were needed to convert a customer order into an approved (engineered) order ready for manufacture. In order to understand the bottlenecks in the process and alleviate them, the company sent individuals out to track specific orders through the processing function to the factory floor. What Northern Telecom found was that paperwork processing, not machines, caused their long lead times. Northern Telcom started to correct the problem areas to make order processing as efficient as manufacturing and to achieve sizable financial and competitive gains in the process. Time and processing steps cut out of the order-administration cycle represent dollar savings as a result of:

- Fewer clerical hours required.
- Improved cash flows.
- Fewer communications dollars spent in trucking and expediting orders (phone charges, faxes, computer time, etc.).

Northern Telecom's efforts paid off when fire destroyed a district telephone switching office in Brooklyn, N.Y., affecting service for thousands of phone users. The phone company went to the largest producers of digital switches to find a replacement unit. Because of emergency conditions, price was not a major consideration. Northern Telecom's bid quoted a delivery time of two weeks—four weeks less than the competitor's. It had even loaded the replacement switching system on a truck before getting approval for the order. Guess which firm received the order?

Acme Boot Company, a division of Farley Industries, is the world's largest manufacturer of Western boots. Until recently it took up to 35 working days from the time a decision was made to manufacture a certain style and size of shoe until the completed batch was delivered to the distribution center. Of those 35 days, actual work was performed for no more than about 14 hours. For 95 percent of the time, then, no value was added to the product (see Figure 10–2).

Acme recently found that about 10 days could be excised by improving the order-processing function and coordinating the ordering of raw materials. This increased the amount of time that value is being added by about 20 percent. More important, it shortened Acme's response time to customers, giving the company the opportunity to sharpen its forecasts and thus reduce finished goods inventory.

LONG-TERM, TIME-BASED COMPETITION RESULTS

What are the ultimate long-term results of a transition to time-based competition? George Stalk claims that "growth rates of three times the industry average, with two times the industry profit margins are exciting—and achievable—targets." Thus, once a company falls behind competitors capable of achieving such results, it may never be able to achieve any sort of industry preeminence.

Stalk describes as typical the example of a firm making equipment for the paper industry. The firm had the largest share of the U.S. market but could not get any more of the business. The reason was delivery times that averaged 22 weeks and were highly variable (plus or minus 10 weeks). Through time-compression efforts, they reduced the lead time to 8 weeks plus or minus 1 day. Volume went up 16 percent and is growing at a projected annual rate of 40 percent. Not only are costs down by more than 5 percent but, with quicker and more consistent lead

FIGURE 10–2
Acme Boot: Idle Time in the Production Process

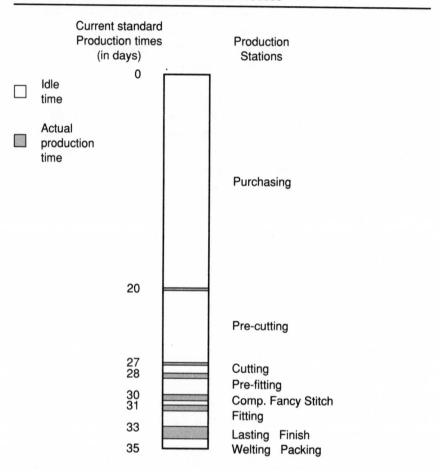

times, prices can also be increased. As a result, profit margins are up by 10 to 15 percent, placing them well above the industry average.

Keys to Becoming a Time-Based Competitor

Becoming a time-based competitor demands major transformations in the way the traditional manufacturing firm is managed. Not surprisingly, these changes are precisely the ones required to convert the manufacturing process itself to a JIT system.

Traditional Manufacturing

Consider the traditional manufacturer. The typical Western manufacturer produces components and products in large batches. Why? To achieve scale economies because the cost of setup, or changeover from one component to another, is so high. In addition, long batch-production runs also contribute to high machine-utilization rates, and this is viewed as superior performance according to conventional accounting standards. However, these large batches lead to large work-in-process (WIP) inventories and, worse, long factory response times.

The traditional manufacturing process is often characterized by a process layout. Processing equipment is grouped according to function, again with the objective of control over machine utilization and efficiency. Product typically follows a slow, disjointed path through such facilities. WIP inventories tend to build up at machine centers, increasing throughput time and making the scheduling task more complex.

To cope with the challenge of a complex scheduling problem, a central scheduling system is typically employed, usually involving a very expensive computer system. Elaborate MRP (materials requirements planning) systems are necessary in this environment to avoid losing track of the large component inventories in the intricate flow process.

JIT Manufacturing

Contrast the traditional manufacturing process with the JIT process. Process batches are small because changeover costs have been minimized. Smaller batches of components yield reduced WIP inventories and dramatically diminished factory response times. By achieving increased product speed through the factory, the transition to a JIT process develops the rapid manufacturing response necessary to be a time-based competitor.

The layout in a JIT factory typically follows product, rather than process, lines. Instead of grouping machines by function, they are grouped into flow lines supporting similar components or products. A simple layout of this type is, of course, very helpful when components are being transferred in small batches. As a side benefit, space requirements are also reduced by sizable amounts (space usually taken up by inventory).

The simpler process flow and more responsive factory minimizes the need for centralized control. Local scheduling rules can be effective in this elemental environment. To schedule production, simple pull systems—"Use one, make one"—can be used. There is less need for a central requirements plan with a complex computer program controlling the flow of components through the plant. As a consequence, the number of dispatchers and expediters should be sharply reduced under local control. With shorter throughput times, forecasting requirements far out into the future is no longer necessary.

None of this provides any surprises for managers who have survived a JIT implementation. What is surprising, however, and is a key to becoming a time-based competitor are the even greater benefits possible by applying JIT principles beyond manufacturing. To see this clearly, examine how these concepts are applied in an entirely different area.

Time-Compressing the New Product Introduction Process

The most positive impact on profits from time-compression activities comes from new-product development (Figure 10–3). A recent McKinsey & Company study of the effects of new product introduction, using these market assumptions—an industry with 20 percent market growth, a five-year product life cycle and 12 percent annual price erosion—found that a six-month delay in entering a market can result in a 33 percent reduction in after-tax profit. To put this in perspective, a six-month delay is five times more costly than a 50 percent development-

FIGURE 10–3
Factors Affecting Profit

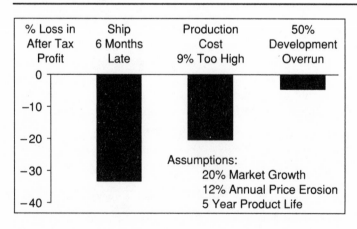

FIGURE 10–4
Who Casts the Biggest Shadow?

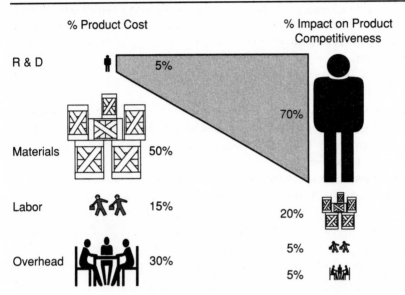

% Product Cost		% Impact on Product Competitiveness
R & D	5%	70%
Materials	50%	
Labor	15%	20%
Overhead	30%	5%
		5%

cost overrun and about 30 percent more costly than having production costs run 10 percent too high.

Getting the jump on competitors by being first to market offers a powerful competitive advantage (see Figure 10–4). The time required to design and engineer a product and then produce it in volume is becoming, in many industries, the key measure of competitiveness.

The Japanese: Some of the best Japanese firms (Honda and Sony, for example) can introduce products twice as fast as their Western counterparts—and have staffs half as large (see Box 1). Coupled with a flexible JIT manufacturing system, these firms clobber the competition with an array of new products at substantially lower costs. This is probably the reason why it is virtually impossible to find a radio, VCR, or CD player designed and manufactured in the United States.

A more detailed review of comparative new-product development data is shown in Figure 10–5. In seven of the eight development steps, the Japanese company took less than half the time the Western company needed. The total difference added up to approximately 18 months; the Japanese got to market 1½ years before the competition.

BOX 10–1
Increased Response Time In Nonmanufacturing Functions

Client Observations

Projection T.V.
- Japanese can develop systems in one-third the time required by U.S. Organizations

Plastic injection molds
- Japanese can develop molds in one-third the time required by U.S. competitors at 30 percent lower cost

Automotive engineering
- Japanese are developing new cars in half the time with half as many people as required by Americans

---

FIGURE 10–5
Improving Response Time in New Product Development

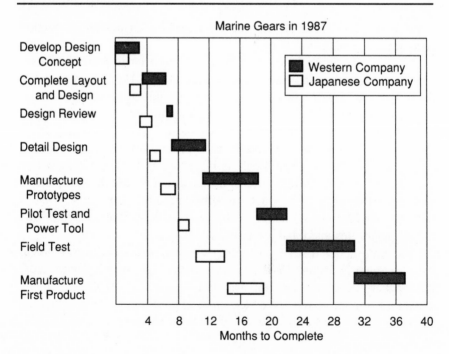

Marine Gears in 1987

Develop Design Concept

Complete Layout and Design

Design Review

Detail Design

Manufacture Prototypes

Pilot Test and Power Tool

Field Test

Manufacture First Product

■ Western Company
□ Japanese Company

4   8   12   16   20   24   28   32   36   40
Months to Complete

The United States: Many U.S. firms are hindered in their efforts to engineer and introduce new products quickly by supply chains accustomed to traditionally slow responses. The following case history illustrates the problem.

A producer of small air compressors in Ohio was introducing a new consumer product. The product concept and price were excellent, but it was crucial to get to market before their major competitor had staked out a claim to that segment of the market. In developing this new product, the firm had two options: (1) produce a prototype for testing at the plant in Ohio, or (2) produce the prototype in Hong Kong and then farm out production elsewhere. The costs of the two options were equal. The decision turned on one fact: The prototype could be produced in half the time in Hong Kong. Lead time to obtain the injection molding dies was over three months from U.S. manufacturers; the response in Hong Kong was, "When do you need it?" The pressures of time-based competition in new product introductions forced the firm to move development and, ultimately, full-scale production offshore.

Leading U.S. high-technology firms have learned the benefits of rapid new-product introductions and have made great strides in improving this part of the process (see Table 10–1). Notably, Hewlett-Packard observed that over half of their sales orders are for products that have been introduced in the last three years. Firms such as Honeywell and Xerox, who have also undertaken major campaigns to increase the pace of new product introductions, have had similar market experiences.

Team Engineering

These firms have time-compressed the new product introduction process

TABLE 10–1
Improvement of New Product Development Times

		Development Time	
Company	New Product	Before	Now
Honda	Automobiles	5 years	3 years
AT&T	Telephones	2 years	1 year
Navistar	Trucks	5 years	2.5 years
Hewlett-Packard	Printers	4.5 years	22 months

Source: *Fortune*, February 13, 1989

by applying the same concepts that enabled their manufacturing arms to adopt JIT. To improve the flow of new ideas, team engineering practices have been introduced. Instead of following a *process* orientation, the objective here is to bring together a team focused on the *product*.

In a team environment, centralized scheduling or monitoring of activities is less necessary. With an automated information process, parallel activities, rather than sequential ones, become the norm— dramatically reducing the time required to develop the new product. Parallel activities, or *simultaneous engineering* as it is known in some circles, also tend to diminish the need for time-consuming product changes and looping back through the development process.

Design Automation
Design automation is employed so that information can be disseminated quickly to all members of the team. This is analogous to the small-batch production so necessary to a JIT process. Design automation means more than buying work stations equipped with CAD (computer-aided design) systems for the engineers. It requires linking all members of the product development team together in a network so that new information can be transferred immediately. Shared databases with information on components and their manufacturability not only speed up the design process; they also yield a product that can be produced more easily, thus compressing the time required for process engineering.

Organizations that Foster Time-Based Competition

What type of organizations are positioned to become time-based competitors? Firms most likely to succeed in a time-based environment are those that are flexible and adaptable and that have top management leadership willing to cut through bureaucratic red tape. Rapid change is required, which dictates a nimble organization. The leaders of such organizations will typically have created an atmosphere of constant improvement and change.

Notably, the strategic plans for two leading U.S. time-based competitors, Hewlett-Packard and Northern Telecom, both employ the term *stretch objectives* to describe plans to reduce time in their processes. Top management typically sets company objectives for the year 2000 (time, cost, profitability, etc.) Using the long-term objectives as targets, they then lay out intermediate goals to be reached over the next six months, year, two years, and so forth. The overall effects of the program are to stretch the capabilities of the entire organization continually over a long time and to keep the firm ahead of the competition.

Flexible Organization Structures

Firms with rigid organizational boundaries will be hindered in their efforts to become time-based competitors. Reducing the cycle time to get products from concept to customer requires teamwork among all the functional areas of the organization. A lack of cooperation among departments can easily frustrate these efforts. Another obstacle is organizational inertia that leads to rigid bureaucratic procedures. People who believe that what they are doing is correct because "we've always done it that way" (another version of the "if it ain't broke . . ." syndrome) are major roadblocks to change.

Teamwork

Organizational structures that support and encourage teamwork seem to be successful, particularly in time-compressing the new product development process. Structures based on function tend to create barriers to team performance. A more flexible organizational structure based on product rather than function will be better suited for time-based competition. In order to speed the process of bringing a new product from concept through engineering to the factory floor, the team must be devoted to the objective of completing the process successfully in the shortest possible time. To be effective, the members of the team must be recruited from different functional areas within the organization. This team cannot perform effectively when the team members must maintain allegiance to their functional leaders. Leading U.S. firms (e.g., Honeywell's Building Controls Division, Xerox, and Hewlett-Packard) have embraced the team concept to spearhead rapid new product introductions. Honeywell has reduced the time required to introduce new thermostats and other control products from 38 months to 14 months with further reductions forthcoming.

Systems Support

Traditional performance measures must be discarded to move into the arena of time competition. Managers are evaluated in terms of time performance rather than traditional cost-based accounting measures.

Machine loadings and overhead allocations are not only outmoded, they are dysfunctional for the time-based competitor: They reward managers for bad decisions. In fact, the same revisions in accounting practices and performance measures required for JIT are needed for time compression throughout the organization.

In designing new systems to support time-based competition, the goal is simple: Focus on time (and quality, of course), measure improvements in the firm's performance along these dimensions, and reward managers on that basis. As with JIT, production speed is of vital importance; things must move and not wait to be counted or batched. Consequently, the importance of individual work centers, around which many accounting systems are based, is diminished. Asset turnover should increase dramatically as the firm becomes more agile and productive, and this should become a key measure.

Information systems play a critical role in time compression by supporting teamwork and coordination across the organization. Rapid exchange of design information among work groups is necessary for the process of simultaneous engineering that is responsible for rapid reductions in new product introduction cycles. Most of the reductions in customer service cycle times have come in the order-processing functions; these changes usually require faster information processing so that orders are processed immediately in batches of one (again applying a basic JIT principle). In fact, much of Federal Express's reputation for reliable, time-specific customer service is attributable to their state-of-the-art information systems. For example, Federal Express now employs a "terminal-in-a-truck" that provides constant contact with the field for tracking packages, expediting deliveries, and planning last-minute pickups.

TIME-BASED COMPETITORS

Toyota

How long does it take to become a time-based competitor? For Toyota, the process appears to be evolutionary and never-ending. Toyota has been refining its production system for more than 20 years and, after making great strides in manufacturing, has now moved successfully to slice the idle time out of the distribution chain.

Photo Processing

At the other extreme are situations where technological changes make dramatic time compression possible virtually overnight. For example, new automated equipment for photo processing has reduced the time required for this service from several days to under an hour. With most manufacturing firms, however, experience indicates that the time-compression process takes years. Implementation must often be achieved incrementally, one department at a time.

Cost is not really a major factor. Most case studies on time-based competition have shown that the competitive benefits far outweigh the costs. Unless costly computer technology is used to speed information flows, most applications involve simple, inexpensive solutions. The time spent planning for necessary changes in a process is often a wiser investment than computerization of every task. Evaluating the costs and benefits of a time-compression program is remarkably similar to observing total quality control (TQC) implementations. With TQC, firms repeatedly report that while the cost of a quality improvement program may appear large at the outset, the process changes that result tend to drive costs down. Time compression, like quality, really is free.

Service Industries

The strategic opportunities of time-based competition are not restricted to manufacturing; time has been the critical dimension of competition in many service industries for years. The fast-food business, of which McDonald's is a leading example, represents the essence of time-based competition. Instead of making customers wait 15 minutes before the lunch arrives, fast-food establishments have food ready and waiting.

The financial services industry also offers significant opportunities. Here the key to gaining a competitive advantage in the marketplace is quick response. The ability to be the first to develop and introduce a new product in response to a change in the tax law is a powerful weapon. Note that the process is remarkably similar to introducing a new product in a manufacturing environment: The financial instrument must be designed, engineered (in the accounting and legal departments), manufactured, marketed, and distributed to the customer. The first firm with a quality product usually gains the lion's share of the market. In the banking industry, for example, we are seeing the emergence of time-based competitors seizing larger shares of the business. By streamlining the process, some banks can now offer 24- or 36-hour mortgages—a process that in the past typically took over a month. Similar changes are sweeping through the commercial loan business.

In the overnight parcel business, Federal Express has used time-based competition to stake out a dominant position while charging premium prices. To maintain that market position, Federal Express has sought to enhance their reputation as a leading time-based competitor. In the past two years alone, they have spent 1.5 billion dollars on capital improvements to make their process more efficient. With this system, Federal Express can handle over a million packages daily and can track down any one of them anywhere in the world within 30 minutes of a request. The time-based strategy has rewarded Federal Express for the past year with revenues of nearly $4 billion and pre-tax profits exceeding $300 million.

In a dramatic recent development, UPS has decided to compete head-to-head on the same dimension—the overnight market. The consumer will be the obvious winner in this confrontation. Marginal competitors in the overnight business will probably be the first casualties.

Identifying the Time-Based Competitor

Firms who have achieved a distinct time advantage should be readily identifiable. Customers are the first to know, so one of the best ways to identify an emerging star is either to ask customers or to assume a customer's vantage point. For the customer, a key barometer is the rate of introduction of new products or technological innovations in current products. Product introductions are a litmus test for time-based competition because this process is the most difficult to speed up. It requires coordination among all functional areas—from R&D and engineering through manufacturing and marketing.

Benetton is a time-based competitor in the fashion apparel business. Through an innovation in the dying process, they dye only finished garments, so production can proceed without a decision on colors until the last minute. If American customers want red sweaters, Benetton can dye more versions red and reduce shipments of other colors. This, of course, helps the company respond more quickly to fashion whims, and it overcomes the disadvantage of the distance between production facilities in Italy and the U.S. market. By being ultra-responsive to the market, the company can keep its investment in finished goods low and increase its profitability.

In the automobile industry, it is easy to spot the time-based competitors. Honda, for example, has one of the highest rates of new product introductions. On the other hand, General Motors has been criticized for a lack of innovation and variety in its product line. With respect to product innovations such as four-wheel steering, for example, U.S. auto manufacturers have not exhibited the rapid product-development capabilities of a time-based competitor.

The computer software market is another area where new product delivery and product upgrades are critical to market success. Spotting the time-based competitor is again relatively easy. Key indicators are the abilities to meet promised delivery dates and to deliver updates quickly whenever new personal computer hardware is introduced. Success flows quickly and surely to the bottom line.

With consumer products, observing how quickly a competitor responds to a product innovation is a useful exercise. When Heinz introduced ketchup in translucent, squeezable plastic bottles, it immediately gained a sizable advantage over its rivals. Caught unprepared, Heinz's competitors have taken a long time to "catch up."

At an operational level a key identifier of an emerging time-based competitor is customer-response time. The manufacturer with a time advantage can promise and deliver orders on short notice. Inventory levels will be lower. In addition, the marketing campaigns of such companies will typically stress their exemplary customer service records.

Equally important is the ability to discern companies that are not time-based competitors. Just ask: Whose products look the same year after year? Whose lead times keep increasing? Whose working capital is tied up in inventories? Whose orders are always late and show few signs of improving? Once these questions are answered, the time-bloated firms become obvious.

Early identification of true time-based competitors is also vital to investors. These are the firms that achieve dominant market positions, lead their industry in profitability, and ultimately outperform stock market indicators. Again, the most successful time-based competitors should already have made great strides in implementing JIT systems (Hewlett-Packard, for example). Many of the firms emerging today as formidable competitors in the arena of time-based competition have already achieved world-class quality and low cost; they have simply added the additional weapon of speed to their arsenal. JIT is an important first step but not the ultimate act of gaining long-term competitive advantage.

NOTES

1. Fredrick W. Smith, "Time: Tomorrow's Weapon for Global Competition," Speech to Time-Based Competition Conference, Vanderbilt University, December 16, 1988.
2. See Goulds Pump case in Chapter 12 for details on compression of order administration time.

CHAPTER 11

TWO TACTICS TO IMPROVE MANUFACTURING PRODUCTIVITY

Richard S. Bombardieri

Editor's note

What practical advice can we offer companies looking to smooth flows and cut down on production chaos? This chapter offers two workable approaches: zero shortages, and continuous work flows. The examples discussed are mining equipment and military electronics, but the principles are applicable to all types of production.

Here are two proven tactics that can be used to support JIT (just-in-time) approaches to improve production operations: (1) Zero Shortages and (2) Continuous Work Flow. These tactics are used to reduce assembly cycle time and WIP (work in process), while concurrently increasing both throughput and yield. Benefits are also realized from reduced staffing. They are very simple, inexpensive tactics that do not require elaborate system support. The following two examples represent typical situations that warrant these tactics.

An electronics buyer purchased a capacitor for half the cost of the one currently in use. The extended savings were large enough for that buyer to receive an award—a certificate and $1,000. Several months later, when the material was placed on the floor for use, the exterior surface was discovered to be too irregular for automated insertion equipment.

Since none of the original components were available, much time
and money was spent inserting the components manually and attempting
to modify the insertion equipment. The results were expensive. Assembly
costs went up; quality and customer service suffered as assembly time
and rework increased. The company was forced to purchase the original
component until a "fix" could be developed. The total cost for the buyer's
attempt at savings was over $22,500.

The lessons learned were that a positive purchase price variance
is not by itself an accurate measurement of cost savings and that shortages disrupt the entire work flow, racking up extra charges along the
way. Further, engineering specifications must consider manufacturing
processes; all changes require thorough communication.

A CASE FOR ZERO SHORTAGES

The second example involves the assembly of large mining equipment.
A primary performance measure for fabrication and assembly was efficiency (actual hours/standard hours). Twenty machines were built per
month, however, so costs for the fabrication area were based on large
lot sizes. Lot sizes were established by traditional EOQ (economic order
quality) calculations and modified by consensus:

"A" item order quantity = 3 months' usage

"B" item order quantity = 6 months' usage

"C" item order quantity = 8 months' usage

In order to reach high efficiency levels during assembly, workers
built with shortages, while expediters attempted to change manufacturing
priorities. Although efficiencies improved to 89 percent from 72 percent,
cost overruns were common. Assembly cycle time was 60 days, inventories were at an all-time high and growing (from 1.4 turns to 1 turn),
and the company was losing money.

Find the Opportunity

Begin by finding the point at which the processes converge and work
backwards looking for bottlenecks. That point is often final assembly
(see Figure 11–1). Why final assembly?

FIGURE 11-1
Walk Backwards through Final Assembly

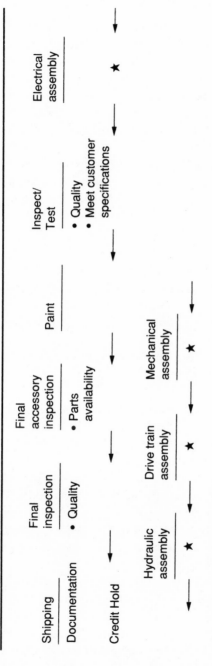

★ Indicates shortages that stop assembly

• Bills of material incomplete or inaccurate
• Parts availability and quality less than 100%
• Assembly documentation and tools incomplete or missing

209

1. The product must be complete prior to shipment.
2. In final assembly it is relatively easy to develop necessary bench-mark statistics for assembly, rework and repair hours, and cycle time.
3. Both quality and delivery for all components can be more easily quantified. The final assembly operation needs 100 percent of the material; it must fit together. Any deviation is obvious and measurable.
4. This area enjoys higher visibility; it represents shipping product. All functions are happy to support any program that will improve their ability to ship on time within cost standards.

Get Started

How to start? Typically everyone understands the problem of assembly shortages, but no one wants to take responsibility for causing them. A *zero shortage* program—simply defined as having 100 percent of all material, engineering, and customer requirements before starting the first assembly operation—is the answer.

The key concept here is *100 percent complete*. MRP (materials requirements planning) systems are designed to release work only when all the identified material is available. People intervene and force the system to release an order that is 90 percent or more complete.

The issue of *relative completeness* is eliminated in a zero shortage program. For example, at a meeting a fabrication manager indicated that an order was 95 percent complete in line items. But assembly could not begin because the frame was one of the shortages. The job was released. The entire work force began expediting. Their charter was to do whatever was necessary to complete the order. Overtime and premium freight charges were authorized, while quality was compromised. The most tangible result, added cost, was accepted as normal. A more dangerous consequence is the unknown impact this practice has on future shipments.

Although the initial application is more easily understood as applied to manufacturing, there are applications for every function throughout the organization.

This program requires the total involvement of everyone in the organization. One individual must be identified to begin the process. He should be accepted by the organization as impartial, should possess good communication skills, and must be highly motivated.

Walk backwards through the operation looking for bottlenecks. Start with the shipping department, work through final inspection and final assembly, and begin looking for problems. At each checkpoint try to observe the operation, and ask the following questions about the process:

Shipping
Are products on the shipping dock because they are missing parts (e.g., software inserted at shipping) or documentation (e.g., a ship-to address, credit hold, or customer-requested consolidation)? *Zero shortages* addresses anything that impacts the ability to ship product.

Final Inspection
The next stop, final inspection, includes both internal and external inspection. Are any products held up for quality reasons, waiting for outside inspectors? Again, both items are required by this department to do its job.

Final Assembly
In final assembly there may be parts shortages, documentation shortages, quality problems, and insufficient tooling or lead time. Identify the cause of the shortages. What department is responsible? Since no single department accepts responsibility, it is possible that everyone is doing his best job as he understands it. This is normal. To understand the issues, start by interviewing each department head, restating the issues in terms that the department managers can relate to and support.

After completing the interviews, schedule a meeting of all department heads to gain acceptance of the issues and the proposed solution. If the homework has been done, there will be senior management commitment to the zero shortage program. At this point, name a multifunctional committee to implement the program.

Implementation Steps

- Define the pilot product or area. Program the project for success by selecting a product that is representative of the issues, is sold on a regular basis, and will benefit by cycle-time reductions achieved with fewer shortages.
- Develop benchmarks (e.g., current cycle time, total labor hours, current WIP level). Clearly identify benchmark objectives, (e.g., 25 percent reduction in cycle time).

- Publish meeting minutes and results. Every event must be examined and reported on bimonthly.
- Develop and publish an implementation plan. Identify specific tasks, schedules, and individuals.
- Identify assembly cycle time, which is when the product is released for picking until it is complete. Total elapsed time is measured, including weekends.
- Develop a best-case assembly cycle. To determine this, divide standard hours by number of assembler hours.

A product requiring 480 standard hours, with three assemblers assigned per shift and two shifts a day is stated as: $\frac{480}{24} \div 2 = 20 \div 2 = 10$ days.

- Determine total labor hours. This is relatively straightforward, but it must include all hours.
- Define standard hours for assembly.

In the earlier case of the mining equipment manufacturer the benchmarks were:

	Standard	Actual Before Zero Shortage Program	After Zero Shortage Program
Assembly Cycle	10 Days	60 Days	20 Days
Direct Labor	480 Hours	650 Hours	515 Hours

Other problems surface in addition to the expected shortages: inventory and stockroom inaccuracies, bill-of-material errors, unresolved order-entry issues, engineering changes, part quality, and missed promise dates.

If bad practices are allowed to infiltrate the daily routine, they become the status quo. In addition to reaping the benefits of implementing a zero shortage program, the stage is set for continuous work flow, the next logical step. The business demands constant improvement. Therefore, reducing the batch size to one and flowing the product forces improvement.

CONTINUOUS WORK FLOW

This concept has proven effective in low-volume military electronics and capital-equipment manufacturing. Again, the benefits result from further reduction of assembly-cycle time, WIP inventory, and rework. In one implementation the business documented indirect labor savings of $700,000. Floor space requirements were cut in half.

Continuous flow is exactly what it says. A job is not started until it can be shipped. Once started, it does not stop or leave the area until it is complete. Additionally, the organization only schedules and makes what is needed for that day. Repairs and rejects are dealt with immediately. The entire organization is challenged to increase effectiveness and to respond immediately to operator needs. Although this concept has been implemented as a stand-alone program, it can only be successful if the released jobs have no shortages.

When the mining equipment manufacturer implemented continuous work flow, cycle time was reduced to 10 days. The assembly process was streamlined. The new assembly standard is now 7 days.

At one military electronics facility the benefits were more dramatic.

	Before	*After*
Assembly cycle	21 days	1 day
Assembly yield	84%	96%
Work-in-process inventory	2700 boards	50 boards
Labor hours	1728	704

How To Start

At this point, floor shortages must have been solved. This enables the organization to concentrate on the process itself from material release to shipping.

If the goal is to release material daily in lot sizes of one, the entire sequence of events, from setting the lot size to the detail flow of that lot, must be understood. The focus should be on the continuous product flow, *not* operation efficiency. Document the actual flow by walking from operation to operation.

Talk to the operators to understand what they are doing, their frustrations, and how they would improve the process.

Improve the Flow

With the flow documented, the committee can now direct each operation and measure its impact on the total product flow. During this process anything is possible; the group should therefore be instructed to ignore any real or imagined resource constraints.

As changes are made and the flow is revised, it becomes easier to define cause and effect in each proposal. This exercise will force the organization to focus on the product process flow as it could be, given the parameters of no shortages, lot sizes of one, and daily work released. It is a back-to-basics approach.

It should not be a surprise to find that the suggested changes do not require large expenditures of capital. Indeed, one business spent more money tearing out conveyors than purchasing capital equipment. The military electronics facility spent less than $200,000, of which $75,000 was spent on consulting to improve their flow. Much of the savings came from reprogramming the existing equipment, not replacing it.

A Continuous Work Flow Application Checklist

Short-Term Opportunities
- Reduce the product structure from multiple levels to one level. Queue, wait, and move times are eliminated. Warehouse receipt and issue transactions are eliminated, and handling damage is reduced.
- Control queues between operations. Set a maximum level for material between operations, and stop production once that level is achieved.
- Cross-train operators. This will allow the operators to help each other as the need arises. Cross-training also permits moving operators periodically to reduce boredom and, more importantly, to introduce new ideas to the process.
- Correct deficiencies immediately. As soon as a part or assembly fails, return it to the offending operation. This limits the number of rejects because the offending operation is notified at once. The operator takes corrective action.

- Move a part or assembly as soon as it is complete. Waiting to accumulate a larger quantity before forwarding does not provide a smooth flow of work. It also increases the number of potential rejects in the system.
- Reduce setup times. Do not accept current times as cast in concrete. Challenge the organization to reduce setups and improve product flow.

Long-Term Opportunities
- Rearrange the area to provide easy handoff between operations. This reduces material handling and improves communications. Supervision can begin to manage by sight.
- Install computer-to-computer communications by using direct numerical control (DNC) versus loading individual tapes. This reduces setup time on autoinsertion and operator-assisted equipment.
- Install help lights at each work station that can only be turned off or on by the operator.

OPERATION EFFECTIVENESS

Improving efficiency is really putting money in the bank. Every worker and operation is challenged to improve. Machine justifications can be made on the basis of improving efficiency.

CASE STUDY: ELECTRONICS FIRM MODIFIES ITS USE OF AUTOINSERTION MACHINE

Changing the focus from *efficiency* to *effectiveness*, one business was able to modify its use of an autoinsertion machine. This piece of equipment had been purchased partly because it could insert more integrated circuits (ICs) faster than an older machine.

The machine was configured to accommodate 60 different ICs for the assembly of very complex and densely populated printed wire board assemblies. Large batches were scheduled, and the machine was programmed to maximize its efficiency.

In any environment, producing at a faster rate than consumption equals excess inventory. This was no exception. To support continuous work flow, the operation needed to be able to handle the worst case — completing each board type in random sequence during the course of one day. To do this while maintaining efficiency standard seemed impossible.

The following questions helped to define required process changes: How many different board assemblies are needed during the course of one day? How many different ICs are required to support these assemblies? If each setup takes 15 minutes, this operation will need 9.5 hours just in setup. This time problem then, would seem to eliminate multiple setups.

A second opinion: One of the project team members could not accept giving up and suggested a radical departure — change the objectives and focus on supporting flow, not machine efficiency. He felt that setup time could be reduced to under one minute. A four-step approach was suggested:

1. Eliminate manually loading the various NC tapes each time a different board is required. To do this, direct wire the machine to the computer holding the original programs in memory. A keyboard is added to the machine, allowing the operator to request a new program (i.e., change the setup).
2. Dedicate specific locations for each IC. Since the machine is capable of handling up to 60 different components, include all 60 in the setup. The others can be handled separately.
3. Reprogram the equipment with the new component locations. The handling tubes must be clearly marked as to specific location.
4. Train the operator in the use and setup of the machine.

Results

The operator demonstrated how quickly and easily the machine could be reprogrammed to accommodate any board type. Total time to change from one assembly to another was reduced to 20 seconds for 80 percent of the assemblies. The other 20 percent required removing and replacing three tube sets of ICs and therefore took additional time — 35 seconds.

The operation could now easily handle the worst case load with time left over. This was so successful that the same concept was transferred to other automated equipment. Total cycle time was reduced to 90 minutes.

This is the same military electronic business that had previously reduced cycle time from 21 days to 1. It is now less than 90 minutes.

A WORD OF CAUTION

A central part of operations planning and control is the application of measurements. Much has been written about how our measurements of efficiency, purchase-price variance, line fill, and economic order quantities, to name a few, are not as relevant as they once were, and in fact, they may contribute to our poor performance. They should not be eliminated, but we need to recognize when they force us into bad habits in an effort to meet measurement criteria.

Measurements of efficiency and utilization can determine how fabrication/assembly operations are scheduled and how purchasing receipts are planned. "My people need more work," is the rallying cry of every shop supervisor. Another fallacy is that we need longer runs or larger buys to reduce costs. We do not have to look very far for examples of how we have implemented these measurements and rewarded personnel who have found the savings, only to find that the "savings" have cost us dearly in both time and money.

CHAPTER 12

JIT FOR WHITE COLLAR WORK—THE REST OF THE STORY

Julie A. Heard

Editor's note

Excellent customer service, responsiveness, and reduction of delivery times are not just manufacturing problems. The front end of our product and service delivery systems needs to be addressed. White collar cells, an understanding of the four cycles (book/bill, design/develop, spec/source, and purchase/produce), and short-cycle management principles will help companies manage better—time is money!

At the conclusion of the chapter, a case study of Goulds Pumps illustrates the practical extension of total-quality principles to the front end of the delivery process, using white collar cells.

To win, not just survive, manufacturing companies must be "best of class" in the market in terms of price, quality, and service. It is apparent that successfully integrating manufacturing and white collar activities is crucial to achieving these objectives. Understanding JIT (just-in-time) is essential to developing the unity of purpose necessary to achieve price, quality, and service objectives.

Unity of purpose is an overriding requirement for success but it cannot be developed until each function understands the three roles it must play. Each must contribute independently to the common purpose.

Each must recognize synergistic opportunities and do its part. Each must avoid sacrificing global benefits to local pressures.

What do the three roles have to do with JIT? JIT provides a conceptual framework to clarify the common purpose and requirements imposed on individual functional areas. It provides a wealth of principles, tools, and techniques that can be used to improve performance in white collar areas as well as in manufacturing. JIT makes it clear when, where, and how each functional area must "pull," "push," or "get out of the way."

AN EXPANDED VIEW

In many markets today, customers have become more discriminating; low product price and high product quality are taken for granted. To gain or even maintain market position in the future, the challenge is to shift the emphasis from the current product-based price, quality, and service variables to superior total enterprise cost, quality, and responsiveness.

To best understand the desires of the customer in today's market, it helps to look at what a typical manufacturing company wants from suppliers. Consider the expectations:

- High-quality, low-cost products.
- On-time order delivery.
- Ability to accommodate reasonable order size and delivery date changes.
- Rapid response to requests for customized or new products.
- Easy order entry and inquiry.
- Correct counts, packaging, and labeling.
- Correct invoicing and appropriate payment terms.
- Fast, accurate, and courteous response to requests for information about pricing, delivery dates, product specs, and the like.

This list is not intended to be all-inclusive. Rather, it is meant to focus attention on all the issues to be considered when rating suppliers. Given the list of what you expect from suppliers, what do you think your customers expect of you? As customers become more demanding, "best of class" takes on new meaning. It refers not just to manufacturing but to everything a manufacturing enterprise does.

THE TOTAL-ENTERPRISE CONCEPT

Today JIT is usually considered a manufacturing philosophy that also provides specific tools and techniques for use in a manufacturing environment. It offers ways to improve product quality and cycle time and to lower product cost. What has been missing is an awareness that JIT applies to the *total enterprise*. JIT involves and includes every function in the company—not just manufacturing. All the support functions required before and after manufacturing activities must also operate.

Extending JIT to the total enterprise is essential. Unless the cost, quality, and cycle-time emphases required of the manufacturing operation extend to support functions, the customers' *perceptions* of the enterprise will likely improve very little.

Extending JIT from the shop floor to the office requires extensive knowledge of its philosophy, tools, and techniques. Everything required to do a job must be in the right place, at the right time, in the right quantity, and at the right quality level. Material (paper and information) must flow. Cycle time must be minimized. The objective must be completed units—not work in process. Continuing emphasis on a "get it right the first time" attitude is crucial. Work must be pulled rather than pushed. "Get ready" must be reduced. User and supplier networks must be created. The philosophy, tools, and techniques of JIT are as applicable to processing paper and information as they are to creating products.

Many manufacturers have used JIT tools to make progress toward lowering product cost and improving product quality and cycle time. A fundamental premise of JIT is continuous improvement. It's time to apply the power of that philosophy to the total enterprise.

THE BUSINESS CYCLES

To achieve the lowest total enterprise cost, highest total enterprise quality, and superior total enterprise responsiveness, manufacturing companies must execute superbly within four major distinct but related cycles (Figure 12–1). In fact, the effectiveness with which an enterprise can execute these four major cycles determines its ability to compete.

FIGURE 12–1
Typical Enterprises

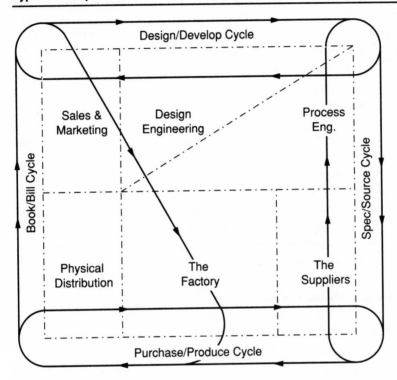

The Book/Bill Cycle

The book/bill cycle varies in length and complexity based on product(s) and market(s). In a grocery store, the cycle is complete when a customer takes an item from the shelf and pays for it. This cycle is substantially longer and more complex in an engineer-to-order environment. No matter which category a company falls into, the length of the cycle has a dramatic effect on customer perception of enterprise responsiveness.

The Design/Develop Cycle

As with the book/bill cycle, the design/develop cycle can vary enormously in length and complexity. Many companies modify standard designs periodically for a variety of reasons ranging from improving ease of manufacture to increasing market share. In addition, many manufacturing enterprises have active campaigns to diversify their product offerings. Others operate almost exclusively in the engineer-to-order market. Performance in the design/develop cycle has an impact on total enterprise cost, responsiveness, and product quality.

The Spec/Source Cycle

During or at the completion of the design/develop cycle, the spec/source cycle begins. This cycle includes all the activities required to bring a product to the point of manufacture: locating sources for commodities, getting parts and tooling and equipment, visiting and qualifying prospective suppliers, holding preliminary negotiations, and so forth. As with the book/bill and design/develop cycles, the activities that make up this cycle require time and cost money. They have an impact, for better or worse, on total enterprise performance.

The Purchase/Produce Cycle

The activities in this cycle are those necessary to acquire raw materials, commodities, or parts needed to produce a product, as well as the actual manufacturing process itself. Activities such as preventive maintenance, materials movement, production and inventory planning and control, and changeover are included in this cycle.

The JIT philosophy has been used effectively in many enterprises to reduce manufacturing-cycle times and improve manufacturing operations. In addition, implementing JIT improves supplier relations by improving incoming quality and reducing material costs. The impact of this cycle on product cost, quality, and customer service is well demonstrated.

TIME IS MONEY

Reducing manufacturing-cycle time is a documented method for improving product quality, reducing product cost, and increasing responsiveness and dependability. However, the cycle-time reduction idea can and should be extended to the total enterprise. A focus on total-enterprise cycle time is important because:

- **All activity takes time and costs money.**
- **The longer it takes, the more it costs!**

Most activities that make up the cycles are performed by white collar workers. Until cycle-time reduction efforts extend to the total enterprise, a great portion of total cycle time within the four cycles will remain unchanged.

Reducing Total Cycle Time

Given the importance of the business cycles to the competitive posture of an enterprise, it is clear that reducing cycle time within each is critical to improving total enterprise effectiveness and competitiveness.

Remember when JIT was introduced in the United States? Changing manufacturing to the extent recommended by the experts was almost inconceivable. Most agreed that changes were necessary; the question was where and how to start. The same "How can I eat that elephant?" mentality is prevalent when considering JIT applications in white collar areas. It is difficult to decide where to start.

When deciding where and how to reduce the total cycle time, much caution should be exercised. Not all improvement efforts are successful. Many fail because they aren't cost effective from a capital requirements or an expense standpoint. Some have negative quality implications. Others require too much implementation time. Still others have multiple thrusts that don't work in concert. To be effective, attempts to reduce total cycle time must improve performance simultaneously on all three dimensions—cost, quality, and responsiveness.

SHORT-CYCLE MANAGEMENT (SCM) PRINCIPLES

With any process of improvement there must be a conceptual foundation to provide a comprehensive and coherent way to attack the problem. Short-cycle management principles (see Figure 12–2) were designed to reduce total enterprise (white collar and manufacturing) cycle time while avoiding potential negative impact. From SCM principles an enterprise can derive strategies, themes, tactical guidelines, and action packages that will improve cost, quality, and responsiveness in everything the enterprise does.

FIGURE 12–2
Short-Cycle Management Principles and Action Steps

Total-Cycle–Time (TCT) Improvement Tactics

TCT tactics are specifically designed to position an enterprise for superior cost, quality, and responsiveness. They apply equally well to white collar and manufacturing-cycle time improvements:

Identify disconnects. This strategy deals with the organizational structure. In most enterprises, there are many disconnects between and among white collar and manufacturing areas responsible for individual tasks in the cycles. This is especially true of the handoffs from a cycle to one or more of the others.

A disconnect occurs when the design department throws a design "over the wall." The design may be producible with current equipment or technology, but tooling may be difficult, time consuming, or expensive to acquire. Disconnects between design and manufacturing engineering and manufacturing make this a common situation.

Remove barriers. This strategy expedites operation of the cycles in an enterprise that has identified and rectified disconnects among and between the white collar and manufacturing areas. It is aimed at the following:

- Policy problems.
- Territorial or priority conflicts.
- Procedural questions.
- Resource needs.
- Authority issues.
- Structural concerns.

Rationalize incentives. Current performance measurement systems often provide incentives for an individual or group to undermine the total enterprise objectives. Rationalizing incentives ensures that everyone works toward the same goal.

An example of irrational incentives is the purchasing agent who is measured on price variance. Low-cost material isn't always the best quality, nor is it always delivered on time. The enterprise objectives aren't served, but that's how current incentives work. If enterprise cost, quality, and responsiveness are objectives, incentives must reward everyone's performance on those dimensions.

Eliminate waste. This strategy concentrates attention on visible and invisible waste in the white collar and manufacturing areas of an enterprise. Waste appears in many forms—work-in-process inventory, defects, unnecessary expenses, excess people, or useless operations, for example. With waste elimination as a strategy, emphasis is placed on a crucial activity.

Manage change. No matter how sound the conceptual foundation or how well-intentioned the improvement sponsors, unless the change process is managed well, the organization will suffer rather than improve. The changes inherent in TCT improvement are massive because they affect the total enterprise. For SCM, activities must be planned, organized, and controlled.

TCT Improvement Themes

As with the tactics, TCT themes place appropriate emphasis on the important white collar activities and issues:

Standardization. To the greatest extent possible, everything should be standardized (methods, procedures, material flows, processes, forms, etc.). Standardization provides task simplicity and task homogeneity. In turn, these foster rapid accumulation of knowledge and experience, thus reducing cycle times for specific tasks.

Autonomy. When restructuring the enterprise, one objective is to create autonomous work teams (white collar cells) with all the resources necessary to operate effectively. Creating autonomous work teams to provide specific services or products focuses the attention of everyone in the team on enterprise cost, quality, and responsiveness. The team takes total responsibility and authority for meeting objectives.

Flexibility. Flexibility in everything is important to reducing total cycle time. From multiskilled workers to variable processing rates to adaptable "structures," flexibility allows immediate response to changing situations and workloads. It lowers the base requirements for an autonomous white collar or manufacturing processing unit (in terms of people, machines, materials, etc.).

Simplicity. In our lust for the most sophisticated planning systems and our quest for high technology, we forgot the beauty of simplicity. Simple systems are easier to teach, learn, and operate. Simple machines are easier to install, troubleshoot, and repair. Simple processes are easier to follow. Simple forms and reports are easier to complete. While searching for better and faster ways, easier and less expensive ways have escaped attention.

Urgency. The excesses in our systems have created a lack of urgency. For example, if a company has six similar machines and one breaks, who cares? But, if one of those machines is linked to four dissimilar machines as part of a process and that machine breaks, the entire process stops. Everybody cares. The same concept applies to people processing paper and/or information. Unless the company structure is focused to create urgency in completing white collar and manufacturing activities, it won't exist.

Visibility. By isolating support staff and technical experts from the "action" and from each other, the visibility of the total process is lost. No one sees the entire scope of activities required to complete one, much less all four, business cycles. It is impossible to solve or prevent problems without an awareness of those problems. Increasing visibility of the total process for the white collar and manufacturing areas of the enterprise is critical to reducing total cycle time.

TCT Improvement Guidelines

The TCT Guidelines provide criteria for evaluating and improving the enterprise structure, as well as white collar and manufacturing operations.

Minimize movement. Movement takes time and requires resources. It often causes delays and confusion. It does not add value. Because white collar and manufacturing processes are highly fragmented, movement is built into current operations. Much movement, both in white collar and manufacturing activities, is waste. When creating plant or office flows, the objective must be eliminating movement to decrease distances traveled and handling required. Reducing movement reduces cycle time more than any other single action.

Maximize focus. Creating autonomous processing units concentrates attention on the important white collar and manufacturing variables—total enterprise cost, quality, and responsiveness. Focusing the enterprise eliminates potentially counterproductive actions that occur when operations are functionally organized. Rigorously focusing an enterprise on a product/service basis rather than organizing work around separate functions (e.g., accounting) allows the visibility required to view, and improve, a white collar or manufacturing process.

Simplify controls. Because many enterprises operate with such complexity, manufacturing and white collar controls on the enterprise are also extensive and complex. By simplifying white collar and manufacturing operations throughout the enterprise, less control is required. By restructuring for maximum product/service focus, the structure controls activities and people. Simplifying controls allows "management by exception" as opposed to the "manage everything" mentality so prevalent today.

Optimize technology. "Technology for the sake of technology" is a common attitude among many enterprises. It prevails in the office and the plant. Sophisticated telephone and computer systems dot the office landscape. Robots, automated storage and retrieval systems, and computer-controlled equipment are found on almost every shop floor. However, high-tech companies are often unknowingly adding needless hours and expense to their white collar and manufacturing processes. Many manufacturing enterprises are currently busy "automating the mess" rather than cleaning it up. Technology can be wonderful; it can replace people in dangerous or tedious jobs. To optimize technology, keep it simple. Use only what's needed for the job at hand. Build it yourself. Exhaust internal ingenuity and brainpower before rushing to the technology store.

Maximize participation. To unleash the true power of the enterprise, everyone must be actively involved in problem solving and TCT improvement. Allow everyone participation in the white collar and manufacturing processes. Everyone has a stake in the competitive posture, thus the continuation, of the enterprise. Let everyone contribute. Listen to suggestions. Record them. Provide feedback. Reward contributions. Empower everyone to help.

In a focused enterprise, doers are team members and are coordinated by the structure. Management creates the structure. Because of the structure, no one can go too far wrong. Have faith in the workers; let them participate in the processes and regard them as their own. The rewards are enormous.

Action Steps

The conceptual foundation provided by the SCM model has been used to improve manufacturing operations and reduce manufacturing-cycle time. In concert with the TCT improvement tactics, themes, and tactical guidelines, the action steps described in this model are universally applicable. The action steps work for both white collar and manufacturing activities (see Figure 12–2).

At the 12 o'clock position in each circle is the business attribute. These attributes are the characteristics that paint a picture of a JIT enterprise. At the 6 o'clock position in each circle is the activating management policy. These policies are commitments that management must make to achieve the business attribute. At the 3 and 9 o'clock positions in each circle are the fundamental concepts related to specific business attributes. They represent the tools and techniques that, if applied in the presence of the relevant activating management policy, allow attainment of the business attribute.[1]

PREREQUISITES FOR WHITE COLLAR JIT

White collar JIT requires smooth, relatively uninterrupted flows. The following example, which is illustrated by Figures 12–3 and 12–4, highlights five areas for improvement:

- Structured flow paths.
- People leverage.
- Continuous flows.
- Linear operation.
- Dependable supply and demand.

It is inconceivable that an enterprise that has attained the SCM attributes is not a JIT operation. It is impossible for any operation to attain JIT status without also implementing JIT principles in white collar areas.

The following example describes a white collar situation containing various flow paths that needs streamlining.[*]

The manufacturing enterprise used for this example builds four distinct products to order (A, B, C, and D). Short Cycle Manufacturing (SCM) has been implemented in the manufacturing operation—each product is built in a separate focused factory. The process for completing the book/bill and purchase/produce cycles for a specific order is as follows:

1. Sales receives and logs an order.
2. The order is transmitted to order entry for paperwork processing.
3. Order entry submits the paperwork to credit for approval.
4. Credit approves or disapproves the order, then transmits the order to P&IC for scheduling.
5. Production Control enters the order on the manufacturing schedule and returns the paperwork to sales for order acknowledgment.
6. Sales acknowledges the order to the customer and transmits the order to traffic for entry on the shipping schedule.

---

[*] Editor's note: The Goulds Pumps case at the end of this chapter illustrates the importance of white collar productivity in customer service and quality improvement. At Goulds, white collar cells review orders; perform tech edits, credit checks, and status inquiries; and work together as a small team. This contrasts with the traditional linear and segregated approach to structuring engineering, customer service, accounting, and order entry functions.

FIGURE 12–3
Before White Collar SCM

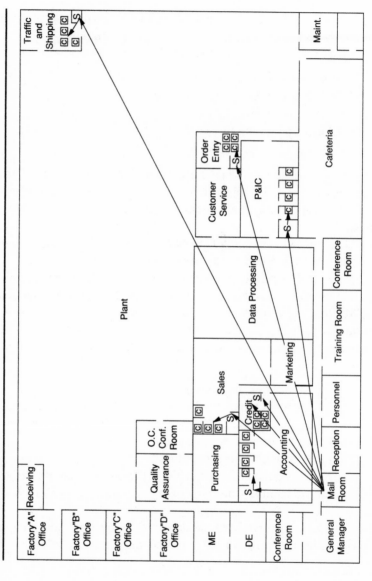

FIGURE 12-4
After White Collar SCM

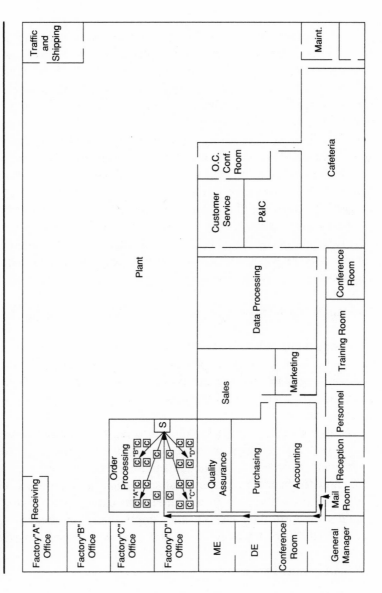

7. After the order is shipped, traffic transmits the paperwork to accounting for invoicing.

Sales, order entry, credit, P&IC, traffic, and accounting are functional departments staffed by white collar workers. By structuring the flow paths differently, efficiencies and reduction of cycle times were realized. Figure 12–3 illustrates the functional layout before JIT practices were applied. All departments were dependent on the mailroom to process their work. The layout was not smooth: Notice how customer service, sales, and accounting all contain separate clerical and supervisory personnel. This organization contains redundancies as well as gaps.

In Figure 12–4, however, the physical layout has been changed to reflect the process, resulting in a smoother, more efficient, more "neighborly" flow. White collar cells have been created in the order-processing area, managed by a single supervisor. The results include decreased processing time, labor cost savings, better communications, and increased responsiveness, since answers are more quickly (and fully) obtained.

Structured Flow Paths

Structuring white collar flows has a specific objective: establishing dedicated, repeatable white collar process paths. These flows eliminate travel time, distance, and handling and avoid confusion and delays. Structuring white collar flows clearly establishes user/supplier relationships in white collar processes.

Numerous user/supplier relationships exist within and between the cycles. Such relationships are often obscure because users and suppliers independently perform tasks that constitute a process. Needs are seldom communicated because of a physical separation between users and suppliers. Feedback occurs even less often. Yet users receive "work" from suppliers, perform their task(s), and supply their "product" to the next user. In fact, every white collar worker is a supplier and a user.

Function versus Process

Structuring (or changing) flows provides other benefits: reduced need for supervision, management, and decision making. In Figure 12–3, workers (indicated by one or more Cs in the departments mentioned) are organized by *function*. Each department requires a supervisor (indicated by an *S* in the departments) to make decisions, distribute and review work, and so forth. Meetings are required to ensure communication between departments, supervisors, and workers.

In Figure 12–4, the structure is based on the *process* (the white collar portions of the book/bill cycle shown in Figure 12–1). The structure determines flows and provides communication among workers— not just among supervisors and managers. No task assignment decisions are required. The process is simple and straightforward. Duplication of effort is visible, so fewer workers are usually required. Workers become more flexible as they watch others work and communicate with colleagues about specific tasks and the total process. The work team is an autonomous unit responsible for the process and has the resources to fulfill its responsibilities.

People Leverage

People leverage means getting employees involved by integrating individual and enterprise goals. Integration occurs when responsibility for cost, quality, and responsiveness is shared. In every enterprise, there are thousands of improvement possibilities and problems. There are, however, limited technical and support staff to solve problems and improve processes.

In a traditional enterprise, people who perform the individual tasks that make up a process are assigned to functional departments isolated from one another by physical and implicit departmental walls. They know the task they perform, but most are ignorant of the rest of the process they're part of. They accept as given that every task performed is required to be performed as dictated. Within the constraints, they perform well. But that's not enough to improve white collar performance.

Process improvement requires that the whole process be visible. In Figure 12–4, the work flows are restructured. All workers can see the entire process. They learn the other tasks involved from one another; that usually leads to questioning the need for some of the steps or finding more effective ways to complete the process. In addition, workers become more flexible as they learn a process rather than just an individual task that's part of the process. Process improvement, however, occurs only if workers have responsibility for the process and authority to improve it.

Continuous Flows

The objective of creating continuous flows is to establish reliability and predictability. Errors, delays, equipment breakdowns, and the like disrupt smooth operation, cost money, and take time. Given an enterprise goal of lowest cost, highest quality, and shortest cycle time in everything undertaken, such disruptions must be eliminated.

The flexibility provided by restructuring work flows and allowing workers responsibility for the process has other implications. They check themselves and each other for accuracy, thus catching errors more quickly. As they discuss the errors, they find ways to prevent them: They failsafe the process. Usually the process is simplified further, thus reducing the likelihood of errors. Process refinement should be inexpensive; allowing the workers to use ingenuity and skill doesn't require outside experts.

Because workers know exactly what to do to complete the process for a single order, a sense of urgency emerges. No longer is one person apt to sit on paperwork. Given the opportunity, workers will standardize the task and the time required to perform it. Process refinement skills are in your enterprise. All you have to do is sanction their use by advocating process refinement and encouraging continuous flows.

Linear Operation

The objective of linear operation is to synchronize white collar activities throughout the enterprise. Creating continuous flows is good; linear operation for white collar activities is better. Linear operation for white collar activities means reducing batch sizes and "get ready" time and establishing daily processing goals.

In Figure 12–3, minimum process time was determined by the flow and the fact that the mailroom only distributes mail twice a day. Given the mail schedule, the minimum single-order process time (if every task were performed in half a day—an unlikely assumption) was three days. With flows restructured to eliminate paperwork movement and handling, minimum process time is dramatically reduced.

Batch size. Independence from the mailroom has a more important implication. No longer do large stacks of paper move from desk to desk and operation to operation. Now, papers move *one at a time*. Batch-size reduction alone improves cycle time dramatically. In manufacturing, reducing cycle time to as near cumulative operation time as possible is the objective. That means eliminating move, wait, and queue time. That objective is appropriate for white collar activities, too, and is accomplished the same way.

Daily goals. Daily goals are set by the process team and their manager (ideally a commitment to complete processing for all orders each day) before team members go home! Establishing daily goals creates the urgency critical to short-cycle management. If the daily work regularly requires a very long day, process refinement comes quickly. People like to go home on time, so they work hard to find more efficient ways of operating, usually by standardizing and simplifying. Linear operation dramatically improves cycle time, thus reducing cost and improving quality.

Dependable Supply and Demand

The objective of dependable supply and demand is to develop partnerships with the antecedent and subsequent operations. Much of the effort to create dependable supply and demand within a process is done by structuring flows, getting people to solve problems to improve the process (for flow continuity), and linear operation. When user/supplier relationships are created in white collar processes, they extend beyond to create user/supplier relationships with other areas within and outside the enterprise. Further development of user/supplier relationships involves customers (and salesmen) and manufacturing operations.

THREE ROLES FOR ALL THE PLAYERS

Implementing white collar JIT principles requires that everyone in an enterprise adopt three roles:

- *Pull.* Adopt total quality control (TQC) and TCT tactics and guidelines to shorten cycle times, reduce costs, and improve competitiveness.
- *Push.* Adopt methods to drive cycle times and costs down and quality, responsiveness, and dependability up in other areas.
- *Get out of the way.* Remove barriers that interfere with TCT efforts in other areas.

The roles are easily understood and appear to be common sense, but filling all three roles every day requires vigilance. It means not buckling under to pressures of the moment. It requires stretching and extending everyone's capability. The organizational structure and culture must foster a willingness to improve perpetually. Everyone must recognize the enterprise goals and commit to them. Everyone must understand the goals and roles before enterprise unity of purpose can exist.

SUMMARY

Every enterprise can and must improve. Price, quality, and service are the *survival objectives*. To attain "best of class" performance with respect to each objective, everyone in an enterprise must understand and apply short-cycle manufacturing management principles. Applying these principles focuses attention on the important competitive variables, develops total enterprise unity of purpose, and provides the conceptual framework, tools, and techniques to improve total enterprise performance.

CASE STUDY: GOULDS PUMPS, OF SENECA FALLS, NEW YORK— TOTAL QUALITY

Patricia E. Moody

How does a 140-year-old company in the industrial heartland go about massive culture change? Goulds Pumps, a very profitable manufacturer of pumps for industry and residential use, is initiating a total quality process along with a total quality project to improve the company's core business processes—distribution, manufacturing, and finance. Although the final results are not yet in, the Goulds' case is important because it represents an example of a company in a mature industry preparing in advance to fend off competition. There are other exciting elements in the Goulds story: organizational structure changes to accommodate the new strategy and the dynamics of the shifts in white collar responsibility.

The basic data describes Goulds' record of profitability in an industry in which margins are tight: 5 percent for those who are doing well. Sales revenues for 1988 were $454 million ($294 million for the Industrial Division), up approximately 18 percent in one year. Goulds is one of the largest pump suppliers in the world. Earnings are up 26 percent, bookings up 15 percent, and return on equity is 13 percent, indicating assets are well utilized.

For five years the demand for basic products (industrial and residential pumps) had been basically flat or down. The market is a tough one in which to compete, and Goulds' strategy was to get through the down period by preparing for the upturn, which started in 1987. The corporate strategy can be summarized as follows:

1. Continue to compete and grow in international markets, which increased 20 percent in 1988.
2. Balance the product lines between commercial, home, and industrial segments; old and new technologies; and domestic and international markets.
3. Invest in the company's future through training for change, accountability, and improved communications.

TOTAL QUALITY, STARTING WITH THE FRONT END

Strategy 3, investing in the future through training and other internal changes, was translated into the drive to create total quality and to improve customer service. Typically, U.S. companies begin quality drives on the production floor. Goulds chose not to. The feeling was that the strategic emphasis should be put on the front end of the system—order processing and customer service. Finished-product quality was not the first priority; instead, top priority was given to improving the company's record of requiring an extended time to process customer orders through technical, financial, and order-entry steps in eight departments.

Order Administration

Thirty-four sales offices prepare handwritten orders that are processed to five manufacturing facilities. The entire process for parts orders (off the shelf) from customer phone call to shipment averaged two weeks;

the new maximum delivery time will be 48 hours. There will be a similar time reduction on pump orders. Management feels that the longer delivery times are symptoms of the system's linearity and that, to offer more responsive delivery times, they must be cut.

The next move is to centralize processing of all parts orders in a new customer service center. The center will serve all customers worldwide (with 800 numbers) and will be open 24 hours per day, seven days per week.

The CATS Project: Competitive Advantage through Simplification

The first total quality project is the CATS project. Through automation and simplification of the order-entry process, the company is realizing reduced costs, faster customer response, and increased sales. (It has been estimated that a 1 percent increase in sales will net $1 million income.) The system will put automation into the field offices. Starting with parts orders, the time to perform the eight-step entry process (quote, credit check, order entry, engineering, scheduling, inventory control, shipping, and billing) is being simplified and compressed.

The payback on the new order-entry system is 18 months. Productivity gains are coming from reduction in administrative costs as fewer people are required to process the same number of customer orders. Personnel shifts to date have realized 10 percent gains—only the beginning of the process change. The other benefits are faster response to customers and increased sales, along with the ability to move people into new jobs more easily.

Performance improvements using the system also include error reduction in pricing, specifications, sales commission calculations, and configurations, as well as speed. In 1988, approximately 198,000 orders were checked manually (even old customers) in the $295 million Industrial Products Group. This averages out to two orders handled per person per day, an expensive and slow process.

The company is looking for improved cash flows from compressed billing lead times; current time from ship to invoice averages three days. With automatic billing, the time will be reduced to less than one day.

Inventory management should improve, too, with better marketing usage data to understand stocking patterns for eight field warehouses. Finished-inventory turns (currently at 4 to 6) are expected to improve.

White Collar Cells

The key change in the way orders are handled involves the creation of white collar cells. The white collar cell is a mix of personnel who are cross-trained to handle all order processing, scheduling, material planning, and engineering tasks. When orders are entered online in the field, time savings are generated as the members of the cell, of which there is one per plant, review the order-entry data and prepare to cut a manufacturing order.

This represents a big culture change for Goulds Pumps as emphasis shifts from individual manual effort to teamwork. The new emphasis is breaking down the barrier between field sales and the plant. Job descriptions within the cells have been broadened to include five specialists per cell. The company has set aside substantial sums for teaching team-building and group dynamics, as well as general total quality awareness. A new position in the field has been created, the District Engineer, a combination customer engineer, customer service manager, and contract administrator.

The Goulds total quality project is an example of culture change in a mature industry. The change strategy is well-defined and well-funded ($3 million in 1989). The commitment derives from the president on down and manifests a very visible impetus toward change. The company is building on the success it has gained competing internationally with high-quality, complex products. The exciting process of reorganizing the front end of a traditional corporate organization into white collar cells will be interesting to watch, especially for other heartland companies for whom change involves less risk than potential for gain.

NOTES

1. For a complete description of the business attributes, activating management policies, and fundamental concepts, refer to "The Direct Route to JIT," authored by Ed Heard and published in the 1986 APICS International Conference Proceedings and "Management Policy Changes—Musts for JIT," authored by Julie Heard, also published in the 1986 APICS Conference Proceedings.

CHAPTER 13

PRODUCTION PLANNING SIMPLICITY

Mark Louis Smith

Editor's note

The following success story shows how one pharmaceutical producer has become more responsive to market demand by cutting production lead times without sacrificing quality or building inventories. The keys to this major change in operations are (1) the empowerment of production workers and (2) scheduling and control systems that are simple, inexpensive, and easily understood by the organization. This chapter describes the planning process from development of the national plan to daily production schedules. The results are impressive.

- Manufacturing lead times have been cut in half not only once, but twice. And, the goal for this facility in the next two years is to cut lead times in half five more times. The final lead-time objective, to make yesterday's sales today, will result in make-to-order manufacturing in a make-to-stock environment.
- Not too surprisingly, over the same five-year period, the work-in-process inventory investment (in dollars not adjusted for inflation) has been reduced by 77 percent. The final objective is to ensure perfect customer service in a volatile, market-driven business.

Does this sound like the saga of a high-technology firm under attack by Pacific Basin countries? Or perhaps a firm that has spent hundreds of millions in capital to completely retool and install supercomputer technology for a lights-out factory? It is neither. It is in fact the oldest manufacturing facility at a leading consumer pharmaceutical company. Further, the primary investment has been directed toward the *simplification* of the production-planning process and the empowerment of its stakeholders, the production workers, to do the scheduling.

HISTORY

This particular manufacturing facility evolved from a small ethical pharmaceutical producer. Over a period of 10 to 15 years, the facility expanded to become the sole production facility of market-leading consumer pharmaceuticals. Today, the facility is one of several highly focused facilities producing multi-ingredient products with low sales volumes and many stockkeeping units.

Ten years ago, the production-planning tools were a collection of homegrown, loosely linked, mainframe computer systems. Today, the tools focus on simplicity and are built around the core of an MRP II system. The critical parts of the MRP II system are its bill of materials (BOM), master production scheduling (MPS), and materials requirement planning (MRP) modules. The remaining critical parts of a production planning system (strategic planning, shop floor control, etc.) have been simplified to the point that personal computers, magnetic scheduling boards, and flipchart paper are the state-of-the-art tools.

THE STRATEGIC PLANNING PROCESS

The planning process begins at the corporate level and ends with daily machine assignments being made by the production operators. Let's look at the four steps required to get to daily scheduling:

1. The National Resource Plan (biannual).
2. The tactical plan (monthly).
3. The rough-cut capacity, material, and finished-goods plans, all plant-level (weekly).
4. Daily production schedules, tracked by shift.

The first step in the planning process is the development of the Five-Year National Resource Plan, published biannually. Corporate planning uses the Five-Year National Resource Plan to develop the 12-month rolling tactical plan. The tactical plan is further broken down into 13 weekly and 9 monthly buckets prior to being passed on to the individual plant planning organizations. At the plant level, the Production and Inventory Control Department develops the master production schedule (MPS), the rough-cut capacity plan, and the material plan, as well as a consolidated finished-goods inventory plan for the plant. The master production schedule is passed on to the individual production centers, where the employees and the production control supervisor develop daily schedules out through material lead times. Execution of the daily schedule is tracked at the end of each shift, with necessary schedule adjustments made by the employees in each production center. Concurrently, the closed loop MRP II system tracks inventory, which feeds back to the plant and corporate planning groups.

Step 1: The National Resource Plan

The Five-Year National Resource Plan (NRP) provides aggregate rough-cut capacities for both capital and skilled labor resources by manufacturing location. It is prepared and reviewed with each manufacturing location twice annually. Local plant management uses the NRP as a template to develop their business and operating plans. The NRP has evolved from a document hand-calculated annually with no simulation capability to today's computer-driven model with the ability to regenerate simulations every 6 to 8 minutes. The model uses simple planning concepts and can be set up on a mainframe using a database language (FOCUS® or QUERY UPDATE®) or on a personal computer using either a database (dBASE® or PFS®) or spreadsheet language (LOTUS 1-2-3® or EXCEL®).

From the NRP, the following strategic operational decisions are made:

- To change a plant's focus, for example, from manufacturing high-volume, single-ingredient products with few stockkeeping units (SKUs), to manufacturing low-volume, multi-ingredient products with many SKUs.

- To change the manufacturing location for a product line in order to eliminate a capacity constraint or to take advantage of a cost differential.
- To make appropriation requests for capital funds to support new products, line extensions, and sales growth and to upgrade or replace aged capital.
- To request approval or justifications for additional skilled-labor sources.

Calculation of the NRP starts with two pieces of information:

- The current annual sales forecast in SKU detail.
- The five-year forecasted growth for each product family (see Table 13–1).

In Table 13–1, for example, the term "Product Family A" represents a group of common SKUs (e.g., 20-, 50-, and 100-count packages) of cough medication.

The next step in the planning process is to develop five years of SKU detail. Each SKU is extended by the forecasted growth for its product family. For instance, in Table 13–1, SKU 1 in Year 2 for Product Family

TABLE 13–1
Five-Year Growth Forecast for Two Product Families

	SKU	Current Year	Year 2	Year 3	Year 4	Year 5
		Product Family A				
Forecasts	1	1,000				
	2	5,000				
	3	3,000				
Total		9,000				
Forecasted growth			5%	4%	4%	4%
		Product Family B				
Forecasts	1	6,000				
	2	4,000				
	3	2,000				
Total		12,000				
Forecasted growth			7%	8%	8%	9%

TABLE 13–2
Five-Year Growth Forecast with SKU Detail

	SKU	Current Year	Year 2	Year 3	Year 4	Year 5
		Product Family A				
Forecasts	1	1,000	1,050	1,092	1,136	1,181
	2	5,000	5,250	5,460	5,678	5,906
	3	3,000	3,150	3,276	3,407	3,543
Total		9,000	9,450	9,828	10,221	10,630
Forecasted growth			5%	4%	4%	4%
		Product Family B				
Forecasts	1	6,000	6,420	6,934	7,488	8,162
	2	4,000	4,280	4,622	4,992	5,441
	3	2,000	2,140	2,311	2,496	2,721
Total		12,000	12,840	13,867	14,977	16,324
Forecasted growth			7%	8%	8%	9%

A is calculated by extending the current year value of 1000 by the year 2 forecasted growth of 5 percent, to yield a value of 1,050 (1,000 × 1.05 = 1,050). In Year 3, the value of SKU 1 for Product Family A is calculated from the year-2 (1,050) value, extended by the forecasted growth of 4 percent, yielding a value of 1,092 (1,050 × 1.04 = 1,092). This process continues for all SKUs and all product families. Table 13–2 shows the completed forecast extensions.

If, however, Marketing or Sales has unique plans for a given SKU, such as a promotional campaign, the extension by growth may be manually overridden.

Explosion of forecasts: Once the forecasts have been developed, they are exploded through the bills-of-material and extended, using standard-machine and labor-planning values in the routings, into machine and labor requirements (see Table 13–3).

Equipment Planning Value: the expected annual output for a specific SKU on a specific production center. Example: Product Family A, SKU 1: annual output = 2,000 units.

Equipment Demand: the amount of equipment needed to support the forecast for a specific SKU (also expressed as machine shifts,

TABLE 13–3
Five-Year Growth Forecast's Implications for Production

	SKU	Current Year	Year 2	Year 3	Year 4	Year 5	Planning Values
		Product Family A					
Forecasts	1	1,000	1,050	1,092	1,136	1,181	
	2	5,000	5,250	5,460	5,678	5,906	
	3	3,000	3,150	3,276	3,407	3,543	
Equipment Demand	1	0.5	0.5	0.5	0.6	0.6	2.000
	2	0.6	0.7	0.7	0.7	0.7	8,000
	3	0.8	0.8	0.8	0.9	0.9	4,000
Total		1.9	2.0	2.0	2.1	2.2	
Labor Demand	1	3	3	3	3	4	6
	2	4	4	4	4	4	6
	3	7	7	7	8	8	9
Total		14	14	14	15	16	
		Product Family B					
Forecasts	1	6,000	6,420	6,934	7,488	8,162	
	2	4,000	4,280	4,622	4,992	5,441	
	3	2,000	2,140	2,311	2,496	2,721	
Equipment Demand	1	0.5	0.6	0.6	0.7	0.7	11,000
	2	0.8	0.9	0.9	1.0	1.1	5,000
	3	0.7	0.7	0.8	0.8	0.9	3,000
Total		2.0	2.2	2.3	2.5	2.7	
Labor Demand	1	3	4	4	4	4	6
	2	5	5	6	6	7	6
	3	6	6	7	7	8	9
Total		14	15	17	17	19	
		Equipment & Labor Totals					
Equipment demand		3.9	4.1	4.4	4.6	5.0	
Equipment supply		5	5	5	5	5	
Loading		78%	82%	87%	93%	99%	
Labor demand		28	29	31	33	35	
Labor supply		86	86	86	86	86	
Variance		58	57	55	53	51	

machine hours, etc.). Example: Product Family A, SKU 1: Forecast = 1000, equipment planning value = 2000, equipment demand = 1000/2000 = 0.5 machines or lines.

Labor Planning Value: the amount of labor required to run the equipment. Example: Product Family A, SKU 1: crew size or labor planning value = 6.

Labor Demand: the average amount of labor required to support the forecast for a specific SKU. Example: Product Family A, SKU 1: Equipment demand = 0.5 machines/annually, labor planning value = 6, average annual demand = (0.5)(6) = 3.

Equipment and Labor Supply are assumed values for the example and would represent the actual resources available.

Following the template provided by the NRP, the Corporate Planning Department and the Plant Production and Inventory Control Departments begin development of the tactical plan.

Step 2: The Tactical Plan (Master Schedule)

The tactical plan is developed at each manufacturing location by the Production and Inventory Control Departments. First, a 12-month rolling forecast is loaded into a traditional MRP II system. Corporate Planning works with Sales and Marketing to develop the forecast. Most of the negotiations revolve around balancing promotions for different product families with available capacities of the manufacturing locations. The goal is to ensure that Manufacturing will be able to provide adequate inventory (to prevent backorder) on a timely basis to support customer service.

Step 3: Capacity, Material, and Finished-Goods Planning

Corporate Planning aggregates each plant's tactical plan to determine the need for temporary transfers of production volume among manufacturing locations to relieve capacity constraints resulting from oversales or seasonality. Additionally, Corporate Planning provides an audit function of the individual plants, keeping an eye out for plan oversights or overextended available resources. Finally, the tactical plan is used to identify excess or dated finished goods. Once these have been identified, Corporate Planning negotiates with Sales on methods for disposition of these inventories.

The tactical planning process, which takes three to four weeks, is repeated monthly.

Step 4: Daily Production Schedules

Now let's take an overview of how the production center develops the daily schedule from the tactical plan. The first 13 weeks of the tactical plan (the MPS) are passed on to the production centers weekly. A Production Control Supervisor (who is a member of the manufacturing staff, not the Production and Inventory Control Department) meets with representatives from each step of the manufacturing sequence and develops the daily schedule through material lead time (typically 3 weeks). The daily schedule is reviewed by Production and Inventory Control. After their sign-off, the schedule is entered into the MRP II system.

While the the theory for developing the daily schedule is not revolutionary, the process is. In the past, the production control supervisor prepared the schedule for the manager's approval. Following approval, the schedule was passed on to the supervisor of each operation. The supervisors assigned wage employees to machines at the beginning of each shift. The employees executed the schedule in more or less the classic American fashion of, "Check your brain at the door and do as you are told."

The process sequence is that of a typical pharmaceutical operation (see Figure 13–1).

The only significant variation is that management is organized vertically; that is, each department has the required assets and bears total responsibility for all steps necessary to produce a grouping of product families. The vertical-slice structure has evolved to eliminate the over-the-fence mentality that existed when each operation was a separate department (e.g., packaging versus processing).

Additionally, by focusing each production center on a small group of product families, manufacturing and marketing have been able to form an alliance. The manufacturing organization can focus on the unique needs (seasonality, launches, etc.) of the different product families.

FIGURE 13–1
Typical Pharmaceutical Process Sequence

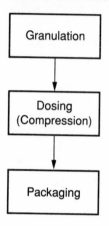

THE NEW SCHEDULING SYSTEM—SIMPLICITY

Today, the production control supervisor meets with wage employees from each operation, who in turn have been empowered to develop the schedule. Once the schedule is developed, it is entered directly into the MRP II system. Although the production center managers do not directly review the schedule, they do monitor customer service (finished goods levels) and key performance indicators (costs, work-in-process levels, and cycle times).

The tools used by the employees to develop the schedule are the tactical plan, a flipchart, a marker pen, a magnetic scheduling board, a calculator, and 15 to 20 years of experience. The type of daily schedule developed by the employees could be generated with a mainframe or personal computer. In the normal course of events, however, the demands placed on production seldom work out as neatly as the computer models. Typically, there are overlapping demands, equipment out of service, production variation, and the like. Because the stakeholders (employees) are responsible for the schedule, they have the freedom to manipulate the schedule to deliver the desired result. The process used is detailed in Figure 13–2.

The process used by the employees is a combination of just-in-time (JIT) and kanban theories. It is JIT in that each operation is scheduled

FIGURE 13–2
Simplified Schedule Process

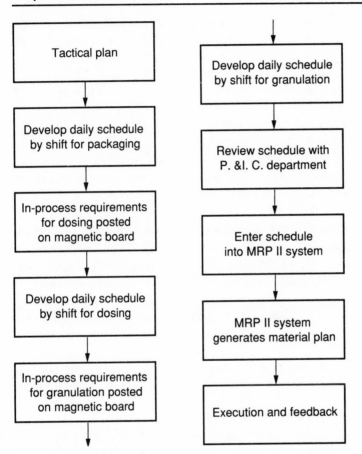

in the smallest possible lot sizes so that the output arrives just when the next operation is ready to begin. Kanban comes into play with the magnetic scheduling board. It is the magnetic tag on the board that signals when each successive upstream operation is to begin manufacturing. Additionally, the scheduling board allows the operators to see the capacity flow of the entire plant.

The employees start the daily scheduling with:

• A sheet of flipchart paper.
• The tactical plan, used to indicate which SKUs are required to be

produced and in what quantities.
• Planning values for each production center.

Armed with this information, the employees assign SKUs to specific production centers. During the scheduling process, the availability of skilled labor is considered and balanced. Needs for unskilled temporary labor are determined and communicated to Personnel. Figure 13–3 shows a completed schedule for a three-shift operation with one skilled crew per shift supporting two packaging lines and four SKUs.

In Figure 13–3, we can see that the production crews are scheduled to alternate between the lines. The cumulative schedule is the amount of production called for in the tactical plan for each SKU (in Figure 13–3, SKU 1 = 72). Finally, each of these schedules would be prepared on separate flip-chart sheets and hung at the appropriate production center at the beginning of the week.

Once a schedule sheet is developed for each machine within an operation, the dates in-process materials are required from each primary department are posted on the magnetic scheduling board. Figure 13–4 shows how the magnetic tags would be posted for the schedules in Figure 13–3.

Figure 13–4 shows that Packaging Line A is calling for the Dosing Department to deliver the materials for SKU 1 at the beginning of the third shift on Monday. A tag is placed on the magnetic scheduling board to indicate the need for the remaining SKUs. Following kanban theory, these tags are the authorization for a primary process to begin manufacturing.

The Packaging Department requirements are used by the Dosing Department to develop their daily schedule sheets. After the Dosing Department finishes preparing their scheduling sheets, they post when they need in-process materials from the Granulation Department. The Granulation Department repeats the cycle of daily schedule-sheet preparation. Figure 13–5 shows the magnetic scheduling board at the end of the process.

Now that the entire schedule has been prepared, all due dates are entered into the MRP II system to generate dock dates for raw materials. Each week the entire process takes approximately 60 minutes to develop the third future week from scratch. Employees rotate the responsibility and work in teams with representatives from each shift to eliminate parochialism.

FIGURE 13-3
Production Crew Flipchart Schedule Sheets

Packaging Line A

Day	Shift	SKU	Schedule	Production	Cumulative Schedule	Cumulative Production	Variance
Mon	3	#1	36		36		
	1		36		72		
	2		Chg. over				
Tue	3						
	1						
	2						
Wed	3	#3	36		36		
	1		36		72		
	2		36		108		
Thr	3		36		144		
	1		Chg. over				
	2						
Fri	3						
	1						
	2						
OT	3						
	1						
	2						

Packaging Line B

Day	Shift	SKU	Schedule	Production	Cumulative Schedule	Cumulative Production	Variance
Mon	3						
	1						
	2						
Tue	3	#2	45		45		
	1		45		90		
	2		45		135		
Wed	3		Chg. over				
	1						
	2						
Thr	3						
	1						
	2						
Fri	3	#4	45		45		
	1		45		90		
	2		Chg. over				
OT	3						
	1						
	2						

FIGURE 13—4
Production's Magnetic Trafficking Board

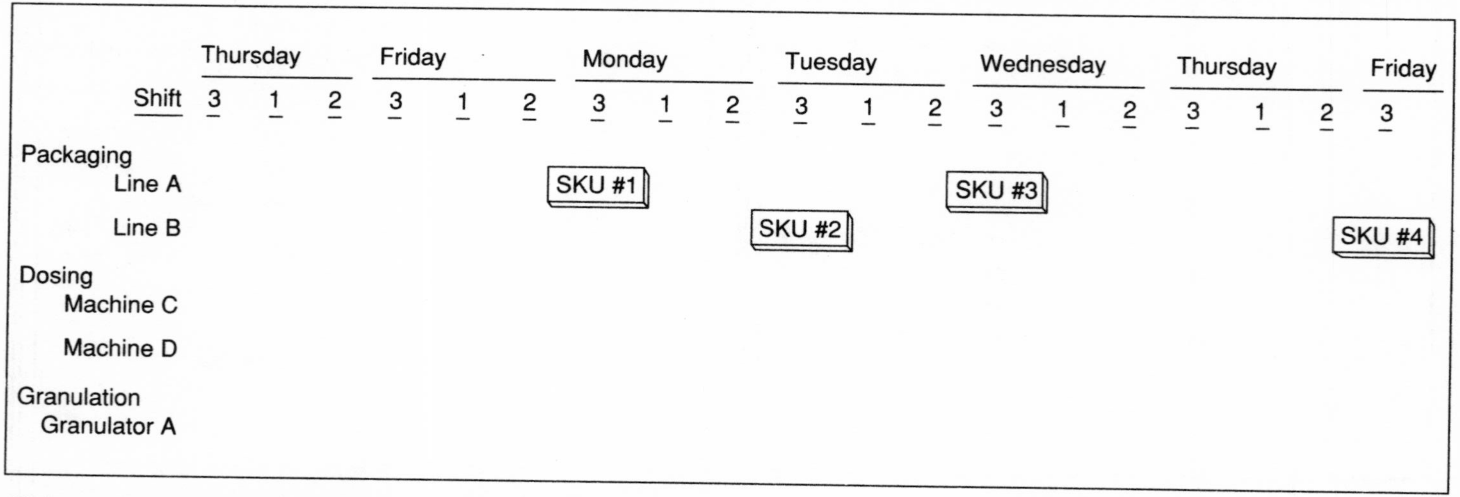

	Thursday			Friday			Monday			Tuesday			Wednesday			Thursday			Friday		
Shift	3	1	2	3	1	2	3	1	2	3	1	2	3	1	2	3	1	2	3	1	2
Packaging																					
Line A							SKU #1												SKU #4		
Line B										SKU #2			SKU #3								
Dosing																					
Machine C																					
Machine D																					
Granulation																					
Granulator A																					

Example Notes: Packaging requirements taken from Figure 13-3.
Leadtime for dosing is four shifts.
Leadtime for granulation is one shift.
All leadtimes calculated from the start of shift.

FIGURE 13-5
Magnetic Trafficking Board at End of Production Process

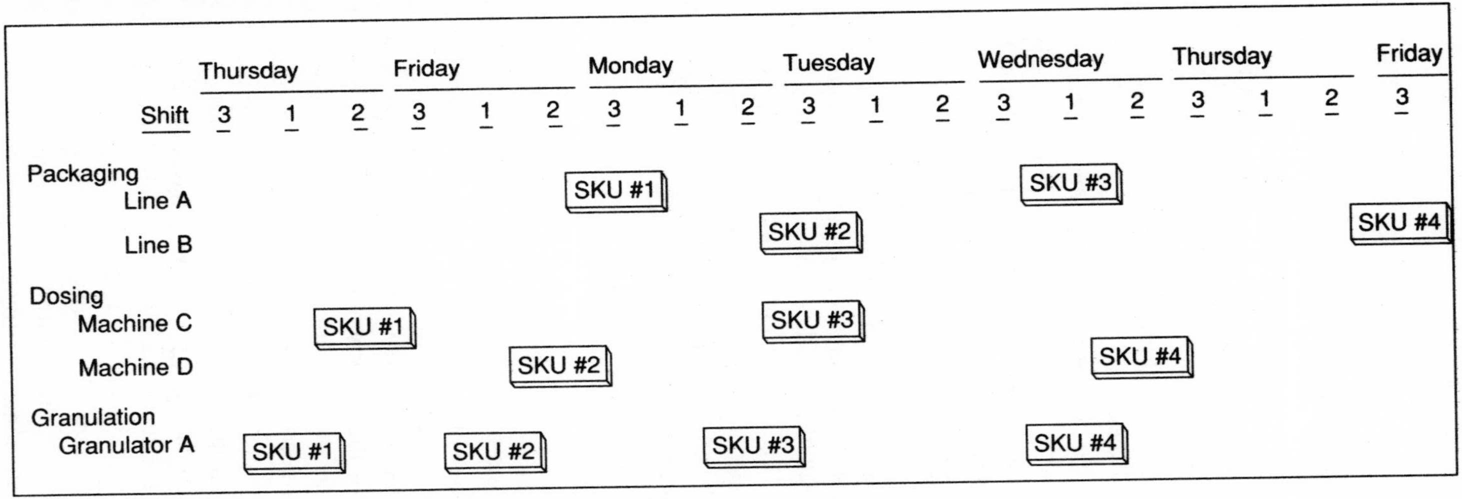

Example Notes: Packaging requirements taken from Figure 13-3.
Leadtime for dosing is four shifts.
Leadtime for granulation is one shift.
All leadtimes calculated from the start of shift.

253

Execution and Tracking

With the production center schedules posted and the material plan in place, all that remains is to track progress against the schedule. Tracking is simply a matter of posting the production at the end of each shift and calculating the variance. Figure 13–6 shows what a typical schedule looks like by midweek.

Packaging Line A is approximately one third of one shift ahead of the schedule (16 units positive variance / 36 units a shift = 0.44 shifts). Meanwhile Packaging Line B is behind the schedule by approximately one third of one shift (14 units negative variance / 45 units a shift = 0.31 shifts). At the beginning of Wednesday's second shift, employees would review the schedule and assign personnel to Production Line B to bring it back on schedule prior to continuing production on Production Line A. Previously, these decisions would have been made by management. If during the week the schedule gets one or more shifts ahead of or behind schedule, the department behind schedule signals the variation to the upstream department by moving the appropriate kanban tag on the magnetic scheduling board. Each upstream department makes the necessary adjustments to its schedules.

CONCLUSION

How is this very simple method different from the old way? Now decisions of daily production rates and mix are made by production workers and monitored by management. Previously, a management decision to adjust the schedule would have been less timely, if it were made at all. The reason for the delay would probably be attributed either to the untimeliness of the variance information or to a preoccupation with the other important aspects of managing a dynamic business.

Although this planning and scheduling method is very simple in its number-crunching aspects (flipcharts, markers and calculators), and in its visible performance-tracking feature (magnets), it requires a shift in energies. Where the hard work before might have been the "mechanics"—software development, user training, generation of hernia reports—now 90 percent of a production manager's job is spent working on *process*. Process in this environment is people management: setting vision and direction, removing roadblocks. Before, a manager had one

FIGURE 13–6
Flipchart Schedule at Mid Week

Packaging Line A

Day	Shift	SKU	Schedule	Production	Cumulative Schedule	Cumulative Production	Variance
Mon	3	#1	36	38	36	38	2
	1		36	35	72	73	1
	2		Chg. over				
Tue	3						
	1						
	2						
Wed	3	#3	36	41	36	41	5
	1		36	42	72	83	11
	2		36	41	108	124	16
Thr	3		36		144	124	
	1		Chg. over				
	2						
Fri	3						
	1						
	2						
OT	3						
	1						
	2						

Packaging Line B

Day	Shift	SKU	Schedule	Production	Cumulative Schedule	Cumulative Production	Variance
Mon	3						
	1						
	2						
Tue	3	#2	45	37	45	37	−8
	1		45	45	90	82	−8
	2		45	39	135	121	−14
Wed	3		Chg. over				
	1						
	2						
Thr	3						
	1						
	2						
Fri	3	#4	45		45		
	1		45		90		
	2		Chg. over				
OT	3						
	1						
	2						

boss; now he works for everyone on the floor. The system requires large doses of communication, meetings, and nudges toward worker empowerment.

For years we have looked at production management hierarchically:

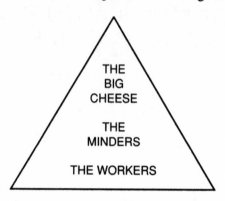

It seems that this system has turned the pyramid upside down.

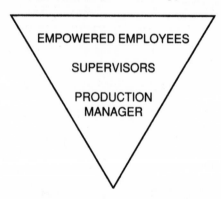

Employees call meetings. They run the business. Life goes on even if shift supervisors are away. In the words of one Production Manager, "I'm 50 percent cheerleader and 50 percent business manager. Some days it would be easier to sit in front of the tube for eight hours, but this is a lot more exciting, as well as profitable!"

CHAPTER 14

NEW MANUFACTURING TECHNOLOGIES

Charles H. Fine

Editor's note

What are the new manufacturing technologies that hold the most promise for U.S. manufacturers? In this chapter Charles Fine looks at the trend toward automation and integration to strategically improve competitiveness on the production floor as well as in support areas. He illustrates this trend with case examples and explores the six management challenges that arise with adoption of these innovative approaches to manufacturing.

INTRODUCTION

Driven by international competition and aided by application of computer technology, manufacturing firms have been pursuing two principal approaches during the 1980s: automation and integration. *Automation* is the substitution of machine for human function; *integration* is the reduction or elimination of buffers between physical or organizational entities. The strategy behind manufacturing firms' application of new automation technologies is multidimensional:

- To liberate human resources for knowledge work.
- To eliminate hazardous or unpleasant jobs.
- To improve product uniformity.
- To reduce costs and variability.

The execution of that strategy has led firms to automate away simple, repetitive, or unpleasant functions in their offices, factories, and laboratories.

When used as an approach to improve quality, cost, and responsiveness to customers, integration requires that firms find ways to reduce physical, temporal, and organizational barriers among various functions. Such buffer reduction has been implemented through elimination of waste, substitution of information for inventory, insertion of computer technology, or some combination of these.

In most *process industries* (oil refining and paper-making, for example) automation and integration have been critical trends for decades. However, in *discrete goods manufacture* (e.g., electronics and automobiles) significant movement in these directions is a recent phenomenon in the United States.

This chapter defines, examines, and illustrates the application of technologies that support the trends toward more automation and integration in discrete goods manufacturing. We begin with a discussion of the technological hardware and software that has been evolving. We then look at six management challenges that must be addressed to support these trends. Finally, we look at the issue of economic evaluation of the new technologies.

AUTOMATION IN MANUFACTURING

As characterized by Toshiba in their OME Works facility, automation in manufacturing can be divided into three categories: factory automation, engineering automation, and planning and control automation. Automation in these three areas can occur independently, but coordination among the three, as pursued by this Toshiba facility, drives opportunities for computer-integrated manufacturing (CIM).

Factory Automation

Although software also plays a critical role, factory automation is typically described by the technological hardware used in manufacturing: robots, numerically controlled (NC) machine tools, and automated material handling systems. Increasingly, these technologies are used in larger, integrated systems, known as manufacturing cells or flexible manufacturing systems (FMSs).

The term *robot* refers to a piece of automated equipment, typically programmable, that can be used for moving material to be worked on (pick and place) or for assembling components into a larger device. Robots also substitute for direct human labor in the use of tools or equipment, as is done, for example, by a painting robot or a welding robot that both positions the welder and welds joints and seams. Robots can vary significantly in complexity, from simple single-axis programmable controllers to sophisticated multi-axis machines with microprocessor control and real-time, closed-loop feedback and adjustment.

A *numerically controlled (NC) machine tool* is a machine tool that can be run by a computer program that directs the machine in its operations. A stand-alone NC machine needs to have the workpieces, tools, and NC programs loaded and unloaded by an operator. However, once an NC machine is running a program on a workpiece, it requires significantly less operator involvement than a manually-operated machine.

A *CNC (computer numerically controlled) machine tool* typically has a small computer dedicated to it so that programs can be developed and stored locally. In addition, some CNC tools have automated parts loading and tool changing. CNC tools typically have real-time, on-line program development capabilities so that operators can implement engineering changes rapidly.

A *DNC (distributed numerically controlled) system* consists of numerous CNC tools linked together by a larger computer system that downloads NC programs to the distributed machine tools. Such a system is necessary for the ultimate integration of parts machining with production planning and scheduling.

Automated inspection of work can also be realized with, for example, vision systems or pressure-sensitive sensors. Inspection work tends to be tedious and prone to errors, especially in very high-volume manufacturing settings, so it is a good candidate for automation. However, automated inspection (especially with diagnosis capability) tends to be very difficult and expensive. This situation, where automated inspection systems are expensive to develop, but human inspection is error prone, demonstrates the value of automated manufacturing systems with very high reliability: In such systems, inspection and test strategies can be developed to exploit the high-reliability features, with the potential to reduce significantly the total cost of manufacture and test.

Automated material-handling systems move workpieces among work centers, storage locations, and shipping points. These systems may include autonomous guided vehicles, conveyor systems, or systems of rails. By connecting separate points in the production system, automated material-handling systems serve an integration function, reducing the time delays between different points in the production process. These systems force process layout designers to depict clearly the path of each workpiece and often make it economical to transport workpieces in small batches, providing the potential for reduced wait times and idleness.

A *flexible manufacturing system (FMS)* is a system that connects automated workstations with a material-handling system to provide a multistage automated manufacturing capability for a wider range of parts than is typically made on a highly automated, nonflexible transfer line. These systems provide flexibility because both the operations performed at each workstation and the routing of parts among workstations can be varied with software controls.

The promise of FMS technology is to provide the capability for flexibility approaching that available in a job shop but with equipment utilization approaching what can be achieved with a transfer line. In fact, an FMS is a technology intermediate to these two extremes, but good management can help in pushing both frontiers simultaneously.

Automated factories can differ significantly with respect to their strategic purpose and impact. Two examples, Matsushita and General Electric, illustrate.

In Osaka, Japan, Matsushita Electric Industrial Company has a plant that produces video cassette recorders (VCRs). The heart of the operation features a highly automated robotic assembly line with over 100 workstations. Except for a number of trouble-shooting operators and process improvement engineers, this line can run with very little human intervention for close to 24 hours per day, turning out any combination of 200 VCR models. As of August 1988, the facility was underutilized; Matsushita was poised to increase production by running the facility more hours per month as demand materialized.

In this situation, the marginal cost of producing more output is very low. Matsushita has effectively created a barrier to entry in the VCR industry by making it very difficult for entrants to compete on price.

The second example is General Electric's Aircraft Engine Group Plant III, in Lynn, Massachusetts. This fully automated plant machines a small set of parts used by the Aircraft Engine Group's assembly plant. In contrast to Matsushita's plant, which provides strategic advantage in the VCR *product market*, the strategic advantage provided by GE's plant seems to address its *labor market*. Plant III's investment is now sunk. Eventually, it will run around the clock at very high utilization rates with a very small crew. As volume is ramped up, GE has the ability to use Plant III's capacity and cost structure as leverage with its unionized labor force, which is currently making many of the parts that could eventually be transferred to Plant III. Thus, factory automation can address a variety of types of strategic needs, from product market considerations to labor market concerns.

Engineering Automation

From analyzing initial concepts to finalizing process plans, engineering functions that precede and support manufacturing are becoming increasingly automated. In many respects, engineering automation is very similar to factory automation; both phenomena can dramatically improve labor productivity, and both increase the proportion of knowledge work for the remaining employees. For many companies, however, the economic payback structure and the justification procedures for the two technologies can be quite different.

This difference between engineering automation and factory automation stems from a difference in the scale economies of the two types of technologies. In many settings, the minimum efficient scale for engineering automation is quite low. Investment in an engineering workstation can often be justified whether or not it is networked and integrated into the larger system. The first-order improvement of the engineer's productivity is sufficient.

For justification of factory automation, the reverse is more frequently the case. The term "island of automation" has come to connote a small investment in factory automation that by itself provides a poor return on investment. Many firms believe that factory automation investments must be well-integrated and widespread in the operation before the strategic benefits of quality, lead time, and flexibility manifest themselves.

Computer-aided design is sometimes used as an umbrella term for computer-aided drafting, computer-aided engineering analysis, and computer-aided process planning. These technologies can be used to automate significant amounts of the drudgery out of engineering design work, allowing engineers to concentrate more of their time and energy on being creative and evaluating a wider range of possible design ideas. For the near future, machines will not typically design products. The design function remains almost completely within the human domain.

Computer-aided engineering allows the user to apply necessary engineering analysis, such as finite-element analysis, to propose designs while they are in the drawing-board stage. This capability can reduce dramatically the need for time-consuming prototype workup and test during the product development period.

Computer-aided process planning helps to automate the manufacturing engineer's work of developing process plans for a product once the product has been designed.

Planning and Control Automation

Planning and control automation is most closely associated with material requirements planning (MRP). Classical MRP develops production plans and schedules by using product bills of materials and production lead times to explode customer orders and demand forecasts netted against current and projected inventory levels. MRP II systems (second-generation MRP) are manufacturing resource planning systems that build on the basic MRP logic, but also include modules for shop floor control, resource requirements planning, inventory analysis, forecasting, purchasing, order processing, cost accounting, and capacity planning in various levels of detail.

The economic considerations for investment in planning and control automation is more similar to that for investment in factory automation than that for engineering automation. The returns from an investment in an MRP II system can be estimated only by analyzing the entire manufacturing operation, as is also the case for factory automation. The integration function of the technology provides a significant portion of the benefits.

INTEGRATION IN MANUFACTURING

Four important movements in the manufacturing arena are pushing the implementation of greater integration in manufacturing:

- Just-in-Time manufacturing (JIT).
- Design for manufacturability (DFM).

- Quality function deployment (QFD).
- Computer-integrated manufacturing (CIM).

Of these, CIM is the only one directly related to new computer technology. JIT and DFM, which are organization management approaches, are not inherently computer oriented and do not rely on any new technological developments. We will look at them briefly here because they are important to the changes that many manufacturing organizations are undertaking and because their integration objectives are consonant with those of CIM.

Just-In-Time Manufacturing (JIT)

JIT embodies the idea of pursuing streamlined, or continuous-flow, production for the manufacture of discrete goods. Central to the philosophy is the idea of reducing manufacturing setup times, variability, inventory buffers, and lead times in the entire production system, from vendors to customers, in order to achieve high product quality (conformity), fast and reliable delivery performance, and low costs.

The reduction of time and inventory buffers between workstations in a factory, and between a vendor and its customers, creates a more integrated production system. People at each work center develop a better awareness of the needs and problems of their predecessors and successors. This awareness, coupled with a cooperative work culture, can help significantly with quality improvement and variability reduction.

Investment in technology—that is, machines and computers—is not required for the implementation of JIT. Rather, JIT is a management technology that relies primarily on persistence in pursuing continuous incremental improvement in manufacturing operations. JIT accomplishes some of the same integration objectives achieved by CIM, without significant capital investment. Just as it is difficult to quantify the costs and benefits of investments in (hard) factory automation, it is also difficult to quantify costs and benefits of a "soft" technology such as JIT. A few recent models have attempted to do such a quantification, but that body of work has not been widely applied.

Design for Manufacturability (DFM)

This approach is sometimes called concurrent design or simultaneous engineering. DFM is a set of concepts related to pursuing closer communication and cooperation among design engineers, process engineers, and manufacturing personnel. In many engineering organizations, traditional product development practice was to have product designers finish their work before process designers could even start theirs. Products developed in such a fashion would inevitably require significant engineering changes as the manufacturing engineers struggled to find a way to produce the product in volume at low cost with high uniformity.

Quality Function Deployment (QFD)

Closely related to DFM is the concept of quality function deployment (QFD), which requires increased communication among product designers, marketing personnel, and the ultimate product users. In many organizations, once an initial product concept was developed, long periods would pass without significant interaction between marketing personnel and the engineering designers. As a result, as the designers confronted a myriad of technical decisions and trade-offs, they would make choices with little marketing or customer input. Such practices often led to long delays in product introduction because redesign work was necessary once the marketing people finally got to see the prototypes. QFD formalizes interaction between marketing and engineering groups throughout the product development cycle, ensuring that design decisions are made with full knowledge of all technical and market trade-off considerations.

Taken together, DFM and QFD promote integration among engineering, marketing, and manufacturing to reduce the total product development cycle and to improve the quality of the product design, as perceived by both the manufacturing organization and the customers who will buy the product.

Like JIT, DFM and QFD are not primarily technological in nature. However, technologies such as computer-aided design can often be utilized as tools for fostering integration of engineering, manufacturing, and marketing. In a sense, such usage can be considered as the application of computer-integrated manufacturing to implement these policy choices.

Computer-Integrated Manufacturing (CIM)

Computer-integrated manufacturing refers to the use of computer technology to link together all functions related to the manufacture of a product. CIM is therefore both an information system and a manufacturing-control system. Because its intent is so all-encompassing, even describing CIM in a meaningful way can be difficult.

We describe briefly one relatively simple conceptual model that covers the principal information needs and flows in a manufacturing firm. The model consists of two types of system components: departments that supply or use information, and processes that transform, combine, or manipulate information in some manner.

The nine departments in the model are:

1. Production.
2. Purchasing.
3. Sales/marketing.
4. Industrial and manufacturing engineering.
5. Product design engineering.
6. Materials management and production planning.
7. Controller/finance/accounting.
8. Plant and corporate management.
9. Quality assurance.

The nine processes that transform, combine, or manipulate information in some manner are:

1. Cost analysis.
2. Inventory analysis.
3. Product line analysis.
4. Quality analysis.
5. Workforce analysis.
6. Master scheduling.
7. Material requirements planning (MRP).
8. Plant and equipment investment.
9. Process design and layout.

To complete the specification of the model for a specific manufacturing system, one must catalog the information flows among the departments and information processes listed above. Such an information-flow map can serve as a conceptual blueprint for CIM design and can aid in visualizing the scope and function of a CIM information system.

Design and implementation of a computer system to link together all of these information suppliers, processors, and users is typically a long, difficult, and expensive task. Such a system must serve the needs of a diverse group of users and must typically bridge a variety of different software and hardware subsystems.

The economic benefits from such a system come from faster and more reliable communication among employees within the organization and the resulting improvements in product quality and lead times.

Since many of the benefits of a CIM system are either intangible or very difficult to quantify, the decision to pursue a CIM program must be based on a long-term, strategic commitment to improve manufacturing capabilities. Traditional return-on-investment evaluation procedures that characterize the decision-making processes of many U.S. manufacturing concerns will not justify the tremendous amount of capital and time required to pursue CIM aggressively. Despite the high cost and uncertainty associated with CIM implementation, most large U.S. manufacturing companies are investing some resources to explore the feasibility of using computerized information systems to integrate the various functions of their organizations.

TECHNOLOGY ADOPTION CONSEQUENCES: FLEXIBILITY AND CAPITAL INTENSIVENESS

As explained above, investments in factory automation and CIM move a firm in the direction of more automation and integration. To fully evaluate such investment opportunities and to weigh the potential payoffs against the costs, one must consider two consequences of these technologies:

1. The flexibility of the manufacturing operation.
2. The capital intensiveness of the operation.

In this section, we look briefly at these two effects before discussing six challenges created by the new manufacturing technologies.

Manufacturing flexibility—the flexibility to change product mix, to change production rate, and to introduce new products—is achieved by shortening lead times within the manufacturing system and by automating setups and changeovers for different products. The importance of manufacturing flexibility for firm competitiveness has become apparent over the past decade as the rate of economic and technological change has accelerated and as many consumer and industrial markets have become increasingly internationalized.[*]

As a consequence of this increased competition, product life cycles shorten as each firm tries to keep up with the new offerings of a larger group of industrial rivals.

To survive, companies must respond quickly and flexibly to competitive threats. Firms must therefore pay particular attention to evaluating the flexibility component of the new manufacturing technologies.

Increased *capital intensiveness* follows directly from automation on a large scale that replaces humans with machines. A transformation to a capital-intensive cost structure has two important effects.

First, the manufacturing cost structure changes from one with low-fixed investment and high unit-variable costs to one with high fixed-investment and low variable costs. This change will significantly affect a firm's ability to weather competitive challenges because low variable costs allow a firm to sustain short-term profitability even in the face of severe price wars.

Second, the changes in both employment levels and work responsibilities brought about by automation require significant organizational adjustment. Challenges brought about by this type of change are discussed below.

SIX CHALLENGES CREATED BY THE NEW MANUFACTURING TECHNOLOGIES

Design and Development of CIM Systems

Because of their ambitious integration objectives, CIM systems will be large, complex information systems. Ideally, the design process should start with the enunciation of the CIM mission, followed by a statement of specific goals and tasks. Such a top-down design approach ensures that the hardware and software components are engineered into a cohesive system.

---

[*]Editor's Note: For more on flexibility, see Chapter 6, "Manufacturing Flexibility: The Next Source of Competitive Advantage," by Sara L. Beckman.

In addition, since the foundation of CIM consists of an integrated central database plus distributed databases, database design is critical. Also, since many people in the organization will be responsible for entering data into the system, they must understand how their functions interact with the entire system. Input from users must be considered at the design stage, and systems for checking database accuracy and integrity must be included.

Hardware and software standardization must also be considered at the system design stage. At many companies, computing and database capabilities have come from a wide variety of vendors whose products are not particularly compatible. Retooling or developing systems to link these computers together requires significant resources.

Obviously, designing a system that will be recognized as a success both inside and outside the organization is a formidable challenge. Few if any companies have fully accomplished this task to date.

Human Resource Management System

As mentioned above, significant adjustment is required for an organization to coalesce behind the implementation of new factory automation and CIM technology. If the new technology is not installed in a greenfield site, layoffs are often one consequence of the change. Reductions in the work force are inevitably associated with morale problems for the remaining employees, who may view the layoffs as a sign of corporate retreat rather than revitalization.

Furthermore, human resource problems are not typically limited to simply laying off a set number of people and then moving forward with the remaining group. CIM and automation technologies place significantly greater skill demands on the organization. Retraining and continuous education must be the rule for firms that hope to be competitive with these technologies; the firm must undergo a cultural transformation.

Requirements for retraining and continuous education are at least as strong for managers and engineers who work with these new technologies as for the factory workers on the plant floor. Designing automated factories, managing automated factories, and designing products for automated factories all require supplemental knowledge and skills not necessarily required for a traditional, labor-intensive plant. Senior managers (who must evaluate CIM technologies), as well as the people who work with them, also can benefit significantly from education about the technologies.

Product Development System

Factory automation and CIM can make product designers' jobs more difficult. Human-driven production systems are infinitely more adaptable than automated manufacturing systems. When designers are setting requirements for a manually built product, they can afford some sloppiness in the specifications, knowing that the human assemblers can either (1) accommodate unexpected machining or assembly problems as they occur or (2) at least recognize problems and communicate them back to the designers for redesign.

In an automated setting, designers cannot rely on the manufacturing system to easily discover and recover from design errors. There are severe limits to the levels of intelligence and adaptability that can be designed into automated manufacturing systems, so product designers must have either intimate knowledge of the manufacturing system or intimate communication with those who do. Developing such a design capability in the organization is a difficult but necessary step for achieving world-class implementation of the manufacturing system.

Managing Dynamic Process Improvement

In most well-run, labor-intensive manufacturing systems, continuous improvement results from a highly motivated workforce that constantly strives to discover better methods for performing its work. In a highly automated factory, there are few workers to observe, test, experiment with, think about, and learn about the system and how to make it better. As a consequence, some observers claim that factory automation will mean the end of the learning curve as an important factor in manufacturing competitiveness. Such an assertion runs counter to a very long history of progress in industrial productivity resulting from a collection of radical technological innovations, each followed by an extensive series of incremental improvements that help perfect the new technology. Most students of the subject estimate that the accumulation of such incremental improvements accounts for as much total productivity growth as do the radical innovations. In essence, any radical innovation may be thought of as a first-pass innovation that requires much more innovation before it reaches its maximum potential.

To presume that factory automation and CIM will reverse this historic pattern is premature at best and potentially very misleading to managers and implementors of these technologies. Because these technologies are so new and so complex, one cannot expect to capture all of the relevant knowledge at the system design stage. If a firm assumes that once it is in place, the technology will not be subject to very much improvement, it will evaluate, design, and manage the system much differently than if it assumes that benefits can be achieved by learning more about how best to use and improve the system once it is in place. One might expect to observe self-fulfilling prophecies in this regard. Even though an automated factory has far fewer people (potential innovators) in it, firms that invest in this technology would be wise to ensure that the people who *are* present are trained to discover, capture, and apply as much new knowledge as possible. In fact, discovering and exploiting opportunities for continuous improvement might be the primary reasons for firms to avoid completely unattended factories.

Technology Procurement

Before evaluating a specific technological option, that option must be reasonably well-defined. A firm needs to choose equipment and software vendors and to decide how much of the design, production, installation, and integration of the technology will be performed with in-house staff. Many observers argue for doing as much technology development in-house as possible to minimize information leaks about the firm's process technology and to ensure a proper fit between the firm's new technology and its existing strategy, people, and capital assets.

For external technology acquisition, technological options must be generated before they can be evaluated. In developing these options, a firm must consider its current assets, environment, and market position, as well as those of its competitors. Equipment vendors must be brought into the decision process. Vendor and technology evaluation criteria must be developed and utilized within the organization.

System Control and Performance Evaluation

Once a technology-investment choice has been implemented, managers typically want to track the efficacy of that investment. The shortcomings of the traditional methods for measuring manufacturing performance are widely recognized. Many of these methods can be manipulated to make current results look good at the expense of potential future results. When managers spend only a small fraction of their careers in one facility or position, they often have an incentive to engage in such manipulations. In addition, in many settings the appropriate performance yardstick for a facility requires information on one or more competitors' facilities for which timely, accurate data may be unavailable.

Increasingly, firms are using multidimensional measures of manufacturing performance. Measures of quality, lead times, cost of quality, delivery performance, and total factor productivity are used to evaluate performance rather than depending on just a profitability summary statistic. Despite this trend, firms could benefit from more research on how, for example, to set standards for productivity and learning rates in a highly automated, integrated environment.

ECONOMIC EVALUATION FOR NEW TECHNOLOGY ADOPTION

The technology adoption costs that are the most visible and easiest to estimate in advance are the up-front capital outlays for purchased hardware, software, and services. Most models consider only these costs. Also important, however, are (1) costs of laying off people whose skills will not be used in the new system, (2) costs of plant disruption caused by the introduction of new technology into an operating facility, and (3) costs of developing the human resources required to design, build, manage, maintain, and operate the new system.

The benefits that flow from investment in factory automation and CIM are both tactical and strategic. These benefits relate to changes in a firm's cost structure, increased process repeatability and product conformance, lower inventories, increased flexibility, and shorter flow and communication lines.

With respect to cost structure, investment in CIM and factory automation tends to represent a large up-front cost that leads to a reduction in variable costs per unit of output. This characteristic results primarily from replacement of labor by machines. Low variable costs can provide significant competitive advantage when interfirm rivalry is high. In addition, reduced variable costs sometimes lead firms to cut prices, potentially increasing market share and revenues.

The advantages arising from the increased repeatability and product conformance afforded by CIM and factory automation can also have significant competitive impact. Decreased process variability reduces scrap and rework costs, a source of variable cost savings that can be as important as the reduction of direct labor costs by automation. In addition, improved product conformance can provide significant sales gains in product markets.

Secondary effects of improved process control include improved morale (and consequent reduced absenteeism and turnover) of employees happy to work in a system that runs well.

Inventory reduction following automation and integration investments can originate from several sources. First, factory automation can reduce setup times for some types of operations, reducing the need for cycle stocks. Second, decreased process variability can decrease uncertainty in the entire manufacturing system, reducing the need for safety stocks. Third, factory integration can shorten manufacturing cycle times, reducing the in-process inventories flowing through the system.

Manufacturing flexibility is another key strategic advantage offered by CIM and factory automation. Rapid tool and equipment changeovers enable firms to change product mix quickly in response to varying market demands. In addition, NC programming and computerized process planning shorten the time to market and time to volume for new products introduced into the factory. Fully automated manufacturing systems provide volume flexibility as well. The highly automated Matsushita VCR factory can change its output rate with relatively low adjustment costs by increasing or decreasing the number of hours it runs each month. Because the factory's direct labor force is quite small, output declines will not lead to dramatic underemployment, and increases do not require major hiring and training efforts.

Finally, reduced lead times between workstations will lower the flow times of work between stations, thus decreasing the need for work-in-process (WIP) in the system. As inventories and lead times are reduced, firms may increase their profit margins by charging more for rapid delivery or may increase market share by offering better service and holding prices constant.

SUMMARY AND CONCLUSIONS

Increased global competition and environmental volatility require that firms adapt quickly or face the possibility of extinction. Investment in automation and integration, including hardware such as automated machines and flexible manufacturing systems, software such as CIM systems, and managerial approaches such as just-in-time and design for manufacturability, can help firms achieve and maintain competitiveness.

Of course, the assets always in shortest supply are managerial vision and leadership. Manufacturing strategy creation must precede technology-investment decisions because good technology rarely saves poor management. Firms must therefore complement their learning about technology options with information and insights about their business challenges and opportunities.

ACKNOWLEDGMENTS

Parts of this chapter have been abstracted, with permission, from the author's paper, "Developments in Manufacturing Technology and Economic Evaluation Models," to be published in *Logistics of Production and Inventory,* S. C. Graves, A. Rinooy Kan, and P. Zipkin (eds.) in the North Holland Series of Handbooks in Operations Research and Management Science. The author gratefully acknowledges helpful comments from Paul Adler, Steve Eppinger, Steve Graves, Patricia Moody, Karl Ulrich, and Larry Wein.

PART 5

MANAGEMENT FOCUS

Part Five, "Management Focus," pulls back from some of the detailed issues of how to execute productivity improvements and emphasizes issues of what manufacturing should and should not be doing in Roger Schmenner's "Seven Deadly Sins of Manufacturing." Frank Leonard's piece addresses how to guarantee that manufacturing and corporate strategies are linked. Ed Heard's chapter has a strong message to deliver to U.S. managers about winning.

CHAPTER 15

THE SEVEN DEADLY SINS
OF MANUFACTURING

Roger W. Schmenner

Editor's note

What are the practical applications of strategic manufacturing theory? The author uses analogy to explain the problems caused by the sins we hear too much about—inflexibility, waste, inconsistency, complication, meandering, impatience, permissiveness, and our favorite, sloth. His manufacturing audit suggests the starting point to identify corporate strategies and align them with production choices.

In the medieval church, there came to be known seven sins that were particularly obnoxious to God: pride, avarice, envy, anger, lust, gluttony and sloth. These were viewed as so heinous, they were termed "deadly" sins, punishable by hellfire.

The medieval church is now history to us but, with the resurgence of manufacturing's importance, it is intriguing to ponder what could send a manufacturing plant straight to hell. Here are my candidates for the seven deadly sins that could do just that: inconsistency, complication, waste, meandering, impatience, permissiveness, and sloth.

INCONSISTENCY

Inconsistency is the candidate for the chief deadly sin of manufacturing. Like pride, inconsistency puts a plant's, or a department's, or a function's interests before the interests of the company as a whole. The economist Kenneth Boulding's phrase, "Sub-optimization is the name of the devil," expresses what is sinful about inconsistency. *Sub-optimization* means consciously, happily, and often scientifically doing what seems right for the plant, department, or function but what in the end, the company may not want or need.

Sub-optimization in manufacturing is often caused by diverse interests tugging and pulling on production people. The needs of the company as a whole get lost in the more immediate needs of segments of the company (finance, marketing, engineering) and what manufacturing can do to further their more specialized interests. With this tugging and pulling comes a host of gripes about manufacturing: Deliveries are not on time, costs are too high, quality is poorer than the competition, and new products take forever to be launched. Manufacturing is regarded as a battleship that takes miles just to turn around. As a result, manufacturing has often felt snubbed by the rest of the corporation.

It is becoming increasingly evident, also, that to become world-class, manufacturing must do many things well. In addition to producing output of consistently high quality on time and at low cost, manufacturing must be flexible enough to accommodate diverse models, fluctuating volumes, and new product introductions. Against world-class competitors, manufacturers no longer have the option of trading these attributes off against one another. We have learned, painfully, that quality and cost are not substitutes for one another but are, instead, complements. This has driven us to consider quality, cost, delivery, and flexibility *all* as complements. This may not be so in every instance, but it is clearly the trend.

Manufacturing Strategy

If manufacturing is to meet the challenges of world-class competition, it has to do so with clear manufacturing strategies, that is, with clear, coordinated decisions about manufacturing's role in the business and how that role ought to be carried out. Manufacturing strategy is getting more attention these days but, regrettably, too often it lies outside corporate strategic thinking. Many thinkers about corporate strategy do not spend a lot of time discussing the intricacies of manufacturing strategy. Rather, they look to manufacturing as a source of competitive advantage solely because of two phenomena: economies of scale and the learning curve. For these strategists, manufacturing benefits a company when it can increase its scale while lowering its cost and when, through experience, manufacturing can move down the learning curve ahead of other companies. What is significant here is that both the economies-of-scale and learning-curve notions are cost related. What the strategists neglect are the many other ways in which manufacturing can be strategically important to the company.

- *The manufacturing strategy process.* Manufacturing strategy is best viewed as an ongoing process rather than as a document. It is a process that feeds off the strategic planning of the company for each of its business units. (See Figure 15–1.) It cannot exist in a vacuum, but must be derived from a business unit's strategy, and it is marketing, not manufacturing, that typically drives that process. Manufacturing's influence, while critical, plays a largely supporting role.
- *Three critical issues for business unit strategy.* Every general manager worth his strategic salt must address three crucial issues. In essence, these issues define what business unit strategy is. The first is *positioning.* The business unit must decide which products it is to produce and what the chief characteristics of those products should be. In short, it must decide which niches in the marketplace it should cover, whether they are just a few or many (a full-line producer). Mercedes Benz, for example, targets only the high end of the automobile market, while GM tries to compete across the board.

FIGURE 15–1
The Manufacturing Strategy Process

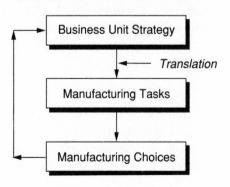

Positioning the company, however, is not enough because change is forever intruding. As long-range plans evolve, the strategy for any business unit is influenced by numerous outside factors that can shape and alter that strategy over time. Product technology is a significant influence on business unit strategy, as are the demographics of the buying population and customers' particular tastes. Competition, too, is a key influence on strategy, as are environmental concerns, including governmental influences on and regulations of business. Coping with these outside factors is crucial to good strategic thinking.

With such constant changes in outside influences, general managers must assess what any *changes in those outside influences* mean for the business unit. This is the second strategic issue general managers must confront. For example, how will the business unit be affected by demographic changes such as the aging of the baby-boom generation, or certain competitor initiatives in particular niches, or government regulation that might affect certain suppliers in the industry. General managers must sort through these influences and determine how the company can be hurt by them or how to win advantage over other competitors as a result of them.

Third, strategists must decide how the company can *exploit* any *advantages* and *minimize* any *disadvantages* that come with these likely scenarios. These are the choices of how the company is to exercise clout in its marketplace, with its customers, and with its suppliers. The examples are endless: Should a company alter the thrust of its advertising to reach older consumers? Should a company seek a distribution channel in China? Should a company oppose pending legislation in Congress? Which of several new technologies seems most promising and worthy of stepped up corporate support? Answers to the questions raised by confronting these key issues are what gives a company strategic direction.

The Manufacturing Task

Unfortunately, business unit strategies are not typically communicated well to manufacturing. The result of the strategy process, the five-year plan, is often reams of paper filled with marketing sales forecasts and financial pro formas. Often there is precious little that manufacturing people can sink their teeth into. The business unit strategy needs to be translated into terms that are meaningful for manufacturing managers. Such a translation is sometimes termed "the manufacturing task."[1] A manufacturing task states what the manufacturing arm of the company *must* do well in order for the business unit strategy to succeed and *also* what it need *not* do well. It sets priorities for manufacturing according to the company's choice of niches it wants to compete in.

Manufacturing Tasks—Three Types

The manufacturing tasks that a company can choose to elevate in importance or to downgrade can be conveniently grouped into three major categories: those that are product related, those that are delivery related, and those that are price related.

1. *Product-related tasks*.   Let us consider product-related tasks first. Four come to mind: performance and features, reliability, new-product introduction, and customization. We can all think about various products and note that some companies typically compete on just one or two product-related aspects, and typically not on all four. For example, some companies are known for their customization: Their products may not include all of the latest new product features that competitors tout. They may not improve performance in many dimensions. They are, however, specific to a particular customer's needs. Likewise, firms that compete on performance and features may not be heavily into new-product introductions or customization.

The auto industry is a readily accessible example. Generally, European cars are noted for their performance and features, Japanese cars for their reliability, while the Big Three American producers have historically competed with customization, allowing consumers to choose from a wide range of colors, interiors, and options. The Japanese, on the other hand, offer very few options. Their cars generally offer a standard package, and while there are a number of options built into this package, individual consumers who may value one option and not another do not have the ability to request a car more customized for them.

2. *Delivery-related tasks*. Delivery-related tasks include the speed of delivery, the reliability of delivery, and what is sometimes called *volume flexibility*. Volume flexibility has to do with how quickly manufacturing can accelerate production and how quickly it can stop production. With many consumer products companies, fads demand a quick ramp-up of production and a quick halt when the fad dies off. Similarly important may be how quickly a product can be made or delivered to the consumer. Quickness is valued by some, but if it is quickness without reliability, some consumers may prefer a more reliable producer, even if that producer may not be the quickest to the marketplace.

3. *Cost to produce.* This manufacturing task, of course, is the one that corporate strategists pick up on first, but low cost is a niche just like any other. In any industry there can only be one, or at most a few, low-cost producers. Others are higher cost, but with other manufacturing tasks done well, they can be every bit as profitable as the low-cost producer. Unfortunately, this truth is overlooked by many, and the result is needless pressure on manufacturing management for lower costs.

Once priorities have been set, there is still a lot of work to do. If manufacturing is going to be successful, all of the numerous choices that manufacturers have to make about the nature of their manufacturing capabilities must be consistent with the manufacturing tasks as they have been translated from the business unit strategy.

Strategic Manufacturing Choices

There are numerous manufacturing choices possible for a company. The Appendix to Chapter 15, "The Manufacturing Audit," lists a number of these choices, divided into three major categories:

- Technology and Facilities.
- Operating Policies.
- Organization.

Consistency in these choices is the hallmark of an effective manufacturer. If the choices are consistent, manufacturing will be a real asset to the company and that strength can be fed back into the business unit strategy. On the other hand, if there are inconsistencies and weaknesses in the manufacturing choices made, then this, too, must be fed back into the business unit strategy to pull the company out of businesses where ordinarily it might want to compete.

Translation: the Toughest Task

The toughest aspect of the manufacturing strategy process is making the translation from business unit strategy to a set of manufacturing tasks. The translation must be in plain language so that manufacturing can discern what its true priorities are and understand how it is to be measured against those priorities. The sad fact of the matter is that the numerous manufacturing choices listed in the Appendix will be made by the company by one means or another. If both manufacturing and general management do not become involved in setting tasks and in making these choices, they will be made by people lower in the organization, including workers and foremen. If companies do not engage in a dialogue about these priorities, they risk inconsistencies across the board. For general management, the question is a very real one: Who do you want to run the company? Do you want general management to run the company, or do you want foremen's and workmen's decisions on strategic matters to hold sway?

As has been argued eloquently by Wickham Skinner in his early work on focus factories, factories cannot be expected to do everything with equal vigor. There must be some choices made, some picking and choosing of exactly what kinds of technologies they are to master, with the understanding that the simpler the operation and its goals and measurements, the better its performance.

Disrupting Consistency

Manufacturing is a lot like a tennis pro who concentrates on a baseline game, where consistency is paramount to winning. Unfortunately, just as in tennis, consistency can be disrupted, and avoiding disruption demands constant vigilance. Companies frequently do not realize how important consistency can be to a factory's performance and, as a result, divergent choices are tolerated, either consciously or unconsciously. In fact, if one gets picky enough about it, one could argue that there is not a factory on the planet that does not suffer at least one inconsistency worth addressing.

In other instances, it is "professionalism," rather than inattention, that disrupts consistency. In most factories, numerous fiefdoms composed of functional work groups are controlled by experts in each function. These people often band together in professional societies, such as APICS, SME, ASQC, and others. While there is much that is admirable about such groups, the dark side is that they encourage overly specialized visions of manufacturing and are subject to enthusiasms out of touch with a particular company's needs. Thus, one runs the risk of having managers push for investments or systems that may be inconsistent with one another. For example, production-planning and inventory-control people may be pushing for a state-of-the-art MRP II system, while manufacturing engineers are enthralled with robots and the quality control department is plugging Taguchi methods. Such initiatives may be totally at odds with what the operation truly needs.

Lastly, change is an enemy of consistency. If things do not change, then merely by trial and error we can devise good manufacturing choices. However, the fact of the matter is that we live in a world of change, and it is change that makes manufacturing more than a trivial exercise.

Product/Process Change Dynamics

To consider the impact on change in manufacturing a bit more, consult Figure 15–2, the product/process matrix, conceived by Robert Hayes and Steven Wheelwright of the Harvard Business School. Hayes and Wheelwright consider the match of a plant's process pattern to the product mix it is responsible for. As Figure 15–2 shows, the process pattern varies from a very jumbled flow, where aspects of the process are only loosely linked together, to the other extreme, where processes are continuous, are usually highly automated, and have a rigidly defined flow to materials. Process segments are tightly linked, one to the other,

FIGURE 15–2
Product/Process Matrix

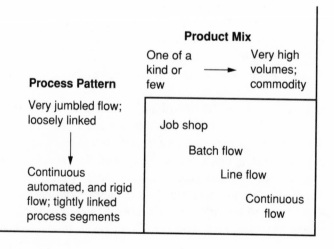

so that one may not even have the sense of moving from one department to another within the process.

There is a continuum for the product mix as well, from one of a kind or, at most, a few units made at any one time to very high-volume commodity products. As can be seen in Figure 15–2, there is a definite correspondence between product mix descriptions and the process pattern that is most effective for making that particular product mix. The diagonal shows the kinds of processes that we have come to identify with certain mixes of product and process. These process categories cut across industry definitions and serve to unify approaches to manufacturing.

Positions off the diagonal carry penalties with them. As can be seen in Figure 15–3, if the process pattern lies more toward the continuous, rigid, and automated process but the product mix stays at high-mix and low-volume levels, the company risks incurring significant out-of-pocket costs. Yamaha attempted to take over the motorcycle market with a high-volume, narrow product line. Honda responded by roaring back with more flexibility and a broader product line. Yamaha was closed out of the high-volume motorcycle market and left with heavy overhead (plant and equipment) costs. The process must be paid for, but the volumes are simply not high enough to pay for the expense of production. This is the quick road to bankruptcy.

On the other side of the diagonal, if the product mix is low and volumes are high but the process pattern is not sufficiently continuous

FIGURE 15–3
The Diagonal and Off Diagonal

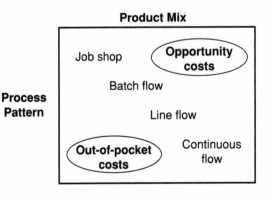

and linked, the plant suffers opportunity costs; that is, the plant forgoes increased contribution and profit by maintaining higher production cost than necessary. This is the slow road to bankruptcy, when the company chokes on the dust kicked by those ahead of it, the low-cost companies that are riding off into the sunset.

These characterizations of "out-of-pocket cost" and "opportunity cost" are important for understanding some dynamics of manufacturing change. Over time, companies frequently experience changes in the product mix that the plant is responsible for. This may often mean that the plant starts out with a low-volume product, perhaps a new product, and as the product catches on in the marketplace, volumes rise. As the product/process matrix tells us, the character of the process should also change to keep the match of product and process on the diagonal. However, it is hard to keep moving precisely along the diagonal. Investments in production tend to be lumpy, and production is seldom changed all at once. This poses a problem.

This problem is not unlike the game of golf (Figure 15–4). The golfer at the tee is faced with the task of hitting a ball onto the fairway, but even Jack Nicklaus cannot count on hitting a perfectly straight drive. Instead, Nicklaus, in his book, *Golf My Way* (New York: Simon & Schuster, 1979), encourages the judicious use of one's natural swing tendencies to hook or to slice the ball. If you naturally hook the ball, aim to the right; if you slice, aim left. There is an important difference

FIGURE 15–4
The Dynamics of Product and Process

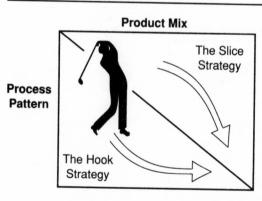

Product Mix

Process
Pattern

The Slice
Strategy

The Hook
Strategy

between a hook and a slice, however. A hook can put the ball farther down the fairway because a hook has overspin on it, and once the ball hits the ground, it can roll far. At the same time, a hook is a riskier shot to take because a ball with overspin can easily roll into the rough.

Because it has backspin on it, the slice typically does not travel as far once it hits the ground and for that reason is less likely to travel into the rough. The slice is thus more conservative than the hook.

Applying the analogy to manufacturing, the hook strategy takes a company into investments in process technology before product volume is developed to the desired level. This is a risky strategy, and it puts pressure on marketing and sales to drum up the volume compatible with the investments in the process. Years ago, Texas Instruments became famous for this kind of preemptive strategy, and more recently the Japanese have been the masters of this kind of approach. On the other hand, the conservative slice strategy is much more popular, but product volume must come first, with pressure on manufacturing to lower costs so that the company can compete effectively with firms that might have already hit farther down the fairway. There are advantages to both strategies, and a company may need to be able to play both hook and slice strategies effectively.

Dealing with such change demands not only the courage to make timely decisions about manufacturing; it also demands attention to both the detail and the consistency required for manufacturing choices to mesh well with manufacturing tasks and with the business unit strategy pursued.

REMAINING CONSISTENT BY
FIGHTING FOR THE FAITH

Weaning a company from the sin of inconsistency demands a good deal of faith. A number of different actions are required that involve more than merely the manufacturing function. Of course, manufacturing can do its part by developing as much internal consistency as it possibly can along the lines emphasized above. This means ensuring that all aspects of operations (technology and facilities, operating policies, organization) are pulling in the same direction. More troublesome are other aspects of management (performance measurement, and cost accounting systems) that can influence manufacturing heavily and can cause inconsistent decisions. Manufacturing thus has to "fight for the faith" to bring these other aspects of management into alignment with its own consistent views.

Manufacturing Performance Measurement

Perhaps the most important battle involves changes in the way manufacturing is measured and evaluated. Current scorecards are often too preoccupied with spotting variances of actual performance from established standards or with focusing on individual cost items within what is presumed to be an accurate accounting of the structure of costs. This promotes reactive, backward-facing management.

The Problem with Cost Accounting

Cost accounting has come under increased fire recently. Even some within the profession itself have asserted that cost accounting is not providing manufacturing with the information it requires in an age filled with rapid technological advances, declining direct labor in the factory, and exploding overheads. Cost accounting is seen as preoccupied with the wrong things: direct labor, measures of labor efficiencies and machine utilizations, and managing to variances of suspect importance.

Solutions

1. *Direct costing.* There have been several suggestions for remedying the situation. Some have called for the adoption of direct-costing techniques that do not assign overhead to specific products but instead focus on contribution margins. Of course, this remedy has been espoused by a small band of followers for years. It has remained a minority position, however, largely because many people want fully allocated product costs for decision making and fear the ambiguity that lies in dealing only with contribution calculations.

2. *Activity-based accounting.* In activity-based accounting, one tries to get a much firmer grip on the real cost drivers for the various overhead accounts. With activity-based cost accounting, overhead allocations that depend only on direct labor or some other simple measure are replaced by numerous, more precise ways of allocating overhead. In this way a more precise cost for a product is developed. While activity-based accounting is conceptually very appealing, many firms are likely

to resist its complexity. For some, the cost of collecting and using the additional information required to allocate overhead more precisely may overwhelm the advantages of such an approach.[2]

3. *Throughput-time accounting.*[3] For those seeking a simple system, yet one without the biases inherent in allocating overhead exclusively according to direct labor, some companies (and my own sentiments) advocate substituting a different rationale as the basis of the allocation: the value of production throughput time. In this instance, the faster a product goes through the factory, the less overhead it picks up per dollar of product value. This introduces an incentive to managers to improve the flow of products in their area because reducing a product's throughput time has the added benefit of reducing its cost, as computed by the cost accounting system.

Furthermore, reducing throughput time may well stimulate a reduction in overhead. How might this occur? Many of the tasks of overhead personnel in a factory are addressed to a minority of products. The old 80/20 rule may apply here, just as it does elsewhere in business; 80 percent of overhead may actually be spent working on just 20 percent of the products. Which 20 percent? My candidates are the 20 percent of products that languish in the factory. These are the products that need more materials handling, more production and inventory control resources, more scheduling, more accounting, more quality control, and more engineering changes. If materials flows are streamlined and throughput times are contracted, some of the time devoted to these overhead activities can be shed.

Finance's Role

Fighting for the faith also means changing the way finance intrudes in manufacturing decisions. The capital-appropriations request process often requires justification for individual investments in new equipment and capacity. Such a mechanism forces attention upon savings readily identifiable to a particular machine, such as reduced direct labor, increased output, or improved quality. This approach leads to justifications for new technology that are more conservative than required. Such an approach ignores investment initiatives that are more broadly based, anchored to a strategy for technological advance and to the potential spillover benefits that one can reap by having several investments linked together. Just because an individual project may not seem to be justifiable does not mean that all the projects, taken together, cannot be justified. Here is where sub-optimization can truly be the name of the devil.

The Role of Marketing and Sales

Fighting for the faith, building consistency, requires a new relationship between manufacturing and the marketing and sales functions. Perhaps the most common cop-out by manufacturing, is that the forecast provided is poor. The forecast is an easy (too easy) whipping boy. In many industries, a disaggregated, item-by-item forecast is very difficult. It makes more sense to improve manufacturing's flexibility and responsiveness rather than to invest more on forecasting. Of course, this means concentrating on lead-time reduction and improving setups and changeovers while at the same time becoming less prey to expediting and firefighting.

Marketing and sales can understand these trade-offs. They can understand that if manufacturing becomes more flexible, they can win new customers, and they can also learn that flagging orders as "hot" or entering orders slowly will only hurt their delivery performance.

Perhaps the most dramatic change that has helped develop consistency in many companies has come in design engineering. Manufacturing is now involved early in the design of the product and to very good advantage.[*] We are seeing now how critical good product design is for manufacturability.

COMPLICATION

The second of manufacturing's deadly sins is complication. Simplicity is virtuous; complication is not. Complication is brought on by both love and fear. Example: there is the love of high technology causing companies to think that they can leap-frog the competition by incorporating new technology in their manufacturing processes. This can happen, of course, but often attempts only complicate what manufacturing should be doing.

Another example of complication derives from the love of increased size. There is scarcely a manufacturer for which bigger does not mean

---

[*] Editor's note: See Chapter 9, "Digital Equipment Corporation: Journeying to Manufacturing Excellence," for design/manufacturing collaboration.

better. Economies of scale carry a real allure, but with increased size comes complication. Often bigger is not really better, considering the complexities that have been added to the process. There are diseconomies of scale: Large plants often suffer from poor layouts and unwieldy bureaucracies.

Finally, love of the system can complicate manufacturing. MRP systems and other computer-based systems have been touted for their ability to simplify manufacturing or to make it more responsive. Yet, in many instances, they lead to nothing more than complication. I am reminded of the manager who observed to me that he had never seen a MRP system that employed fewer people once it was implemented. We embrace systems, but they are often band-aids. We ignore the real possibility of rethinking and simplifying the process and the information flow that supports it.

Complication is also brought on by fear, such as fear of messing up a timetable. Managers do not want to be responsible for moving back the date of anything, even if it means living with complication. Taking some time up-front may be what is truly called for, but it is an option easily passed over. Then there is fear of managing customers or of managing suppliers. Some manufacturing complications occur because managers are unwilling to take the bull by the horns to alter the way customers or suppliers operate, even when such alterations might, in fact, be welcomed by customers and suppliers both.

Finally, complication is often not recognized for what it is. Rationales that I have heard too often are: "We may be more complicated in these operations, but our operations are different." "We serve a different market." "We have a different process." "We have particular constraints that are different from others." We have learned through the years, however, that manufacturing firms share many common traits. Most companies are not different. Basically, they are really very much the same. If we recognize that, maybe we can spend some time eliminating the complication in manufacturing. It may just mean not going with high technology, remaining small, developing simple systems, sacrificing the timetable, and managing customers and suppliers.

WASTE

The third deadly sin is waste. Anyone familiar with total quality control (TQC) and the just-in-time (JIT) manufacturing movement is familiar with waste as a sin. As Shigeo Shingo has pointed out, waste comes in many forms: overproduction (one more than necessary is as bad as one less than required), inventories of any size, waiting for work to be done, extraneous motion in doing the job, transportation between operations, defects that may generate rework or scrap, and processing itself (Does the operation have to be done at all?). One can go far reducing cost and improving upon other manufacturing tasks solely by concentrating on the reduction of waste, but diligence is required. One must go beyond the simple implementation of statistical quality control (SQC) techniques and on to problem solving. One must even go beyond the factory floor to push SQC and problem solving into the overhead indirect labor, such as order-entry and customer service, functions.

Value-Adding Activities

Indeed, what needs to be done is for more of the factory's time to be devoted to value-adding activities. Still another 80/20 rule may apply. In many companies, of all the activities actually performed, only about 20 percent of them typically involve value-adding steps, where a part is actually machined or painted or assembled. Eighty percent of the steps along the way do not add value to the product: counts, inspections, movement, handling of paper work. When we can remove the steps that do not add value, we remove waste.

Throughput-Time Reduction

The most effective strategy for pushing waste aside is concentration on throughput-time reduction.[4] We have already alluded to this in discussing cost accounting. Throughput-time reduction can promote so many good things: quality improvements, lower inventory levels, rapid attention to bottlenecks, and diminished confusion in the factory caused by such things as engineering change orders or expediting. As mentioned in Solution 3 of the section, "Manufacturing Performance Measurement," throughput-time reduction can also promote fewer transactions in the factory and thus less overhead.

In the pharmaceutical industry, Eli Lilly's efforts in this area have paid off. Production-process time has been reduced from between 45 and 60 days to 6 days. Harley Davidson implemented changes that resulted in the following four improvements in their key indicators:

- Inventory turns went from 4 or 5 to between 19 and 23.
- Productivity increased 55 percent.
- Scrap and rework decreased 80 percent.
- Throughput time was cut from 72 days to 1 day.

These improvements cumulatively allowed the company to increase marketshare in heavy motorcycles by 10 percentage points.

Concentration on throughput-time reduction can begin to question all sorts of transactions that take time and do not directly add value to the product: counts, inspections, materials handling, monitoring of direct labor, the tracking of work-in-process inventory. Indeed, anything that generates paperwork is suspect.

Perhaps the most appealing aspect of throughput-time reduction is that it can mean something to everyone, and while that "something" is often different for different people, it can be useful for manufacturing performance. To the worker on the floor, for example, throughput-time reduction can be applied to work flowing through his workstation. It can promote improvement in the layout of the workstation itself, better materials handling, and more efficient setups and maintenance. To the department manager, concentrating on the reduction of throughput time can mean something else, such as improving the design and linking of workstations, lowering the level of inventory, and stabilizing the mix of products made. To the plant manager, reducing throughput time may mean altering the designation of what departments are responsible for and how managers are evaluated—for example, by moving to the identification of product flows that use designated equipment and are evaluated separately. To the corporate VP for manufacturing, reducing throughput time can involve multiplant strategies. What is important is that, in each case, reducing throughput time can be both understandable and helpful in the crusade to eliminate waste.

MEANDERING

Meandering, the fourth deadly sin, defined as "wandering aimlessly," has dramatic consequences for manufacturing. Bad layouts, for example, have no logical flow; they wander aimlessly. It is too easy to leave a layout alone, to reorganize it as expansion occurs or as products proliferate. Failure to move machines and people into a more organized layout has deleterious implications for a factory's competitiveness.

Meandering also means neglecting to develop the family of common parts. It is failure to see what may be common about the operation and what is truly different. Organizing for the smooth flow of a family of parts is often a more efficient way of running the business.

Chasing low-cost direct labor all over the globe is another form of meandering. In deciding where to produce a particular product or component item, current practice in many companies compares fully burdened direct labor costs in different countries. The decision disregards all the indirect and intangible costs to manufacturing competitiveness and flexibility caused by moving materials around the world. While the cost-accounting system may say that sourcing in the Far East or in Latin America is the way to reduce product costs, the aimless wandering that decision implies for the product flow is something to guard against.

Meandering can also mean that products or processes will be assigned to plants solely for the reason that space is available. Just as nature abhors a vacuum, top management abhors vacant manufacturing space. The careless assignment of products to plants causes aimless wandering of materials, technology, and management effort.

IMPATIENCE

The fifth deadly sin is impatience. Of course, there are many companies that value impatience, or urgency, in their managers. "Sinful" impatience refers to such evils as expediting and work-around (releasing an order— without all the components on hand—with instructions to "work around" the components not yet there). These are misdirected rushes in the factory that interfere with manufacturing's abilities to produce quality with timely delivery. Expediting and work-around ensure that the company engages in additional overhead to cope with such situations and adds to the time it takes products to go through the process.

Engineering change orders (ECOs) are another source of impatience in many companies. Many managers feel that if engineering changes are not made right away, sales will be lost because the design is not the best it could be. Yet, how many engineering changes are truly necessary, especially right away? Many companies are willing to dribble out a never ending stream of engineering change orders, and in some cases an ECO exists only to correct a previous ECO that was released prematurely. The fact that it may be easy to release an engineering change order does not mean that it should be done. The process can be abused. A more patient, more deliberate approach to the matter, perhaps bunching engineering changes together, can be advantageous.

Impatience is also reflected in a company's attitude toward quality improvement. There is a difference between instant improvement and continuous improvement. Impatience can focus people on the quick fix and away from the habit of continual improvement that the company must have to be a competitor over the long term.

Lastly, there has been much written about the short-sightedness of the typical American manager, his preoccupation with the quarterly earnings report, and his willingness to forgo long-term benefits for short-term goals. It is hard to quantify this attitude. We have anecdotes but precious little hard research. However, we know the destructive effect of short-term thinking on organizations.

PERMISSIVENESS AND THE PARADOX OF FLEXIBILITY

Permissiveness is perhaps the hardest sin to comprehend, but it is firmly linked to flexibility, a key manufacturing task. *Permissiveness is a sin because it breeds rigidity in the process*. This seems paradoxical. How does it occur? When the process is permissive—when rescheduling is done all the time, when expediting is permitted, when engineering change orders are rampant—paralysis sets in. The process is unable to put anything out on time and loses its grip on quality. As has often been observed, when everything in the factory is tagged red for immediate action, that's as good as saying nothing is tagged red.

Conversely, *discipline breeds flexibility*. This is the paradox of flexibility. Consider the job shop. A job shop is tremendously flexible in the mix of products that it can produce: large quantities, small quantities, all with different characteristics. However, a job shop only operates well if it has an absolutely rigid flow to the information and recordkeeping within it. Tooling, setups, and the load on particular workstations can change dramatically in a job shop, but the flow of information (the paperwork, the recordkeeping) has to run like clockwork or the job shop becomes hopelessly snarled.

Consider another paradox, one related to quick change setups. Quick change setups are the quintessence of flexibility. But think what it takes to be able to change setups quickly: prescribed methods for making them, similar tooling designs, well-conceived workstation layouts, and strict machine maintenance schedules and procedures. Discipline breeds flexibility.

Flexibility Types

There are, of course, different kinds of flexibility that one may want to foster. There is the product mix flexibility that is best characterized by the job shop. There is also the flexibility of new-product introduction, being able to add products easily and keep up with the state of the art. There is also volume flexibility, being able to ramp up production quickly or, alternatively, to cut it off quickly in order to capitalize on fads or to react to severe changes in marketplace conditions. These different types of flexibility demand different kinds of disciplines.

- *Product mix flexibility.* Product mix flexibility requires quick changeovers, and, as mentioned above, quick changeovers demand their own rigidities. Product mix flexibility requires a cross-trained workforce, able to operate several machines simultaneously, able to move from one bottleneck operation to another. Freezing production schedules is helpful, and expediting is not permitted. Product mix flexibility also demands discipline in the sharing of information, both within the factory, with the sales staff, and with vendors.
- *New-product flexibility.* New-product introduction, on the other hand, requires a different set of disciplines. The companies that do new-product introductions best (Digital and Hewlett-Packard, for example) make sure that a rigid set of reviews and sign-off procedures is adhered to at various stages. In addition, there may be procedures through which all the engineering drawings must pass, as well as procedures and people through which engineering changes must go. Nothing falls between the cracks. Good new-product introduction typically requires engineering follow-up. Engineers know in advance that they are to be responsible not only for the design, but for initial production as well.
- *Volume flexibility.* Volume flexibility requires still different disciplines. The most important is maintaining a cushion of capacity resisting any management wishes to see it fill up with unnecessary product. Without a cushion of capacity, manufacturing has

great difficulty increasing production quickly. Volume flexibility is enhanced by general-purpose, convertible equipment; by layouts that are easily changed; by subcontracting of jobs; by temporary hires; and by cross-trained workers willing to move from one task to another.

Effectiveness in volume flexibility may also require low breakeven levels derived from a more labor-intensive operation than found in plants that do not value volume flexibility.

Winnebago and Chrysler have both mastered the art of building volume flexibility. Winnebago's recreational vehicle business is subject to demand swings caused by gasoline price fluctuations. Yet the company knows how, when necessary, to cut overhead and exist on very low volumes to survive. Winnebago has also demonstrated an ability to gear up quickly.

Chrysler has learned to effectively downsize. When faced with disaster, the company's restructured operations produced lower breakeven points.

The Paradox of Flexibility

The paradox of flexibility is this: the extent and nature of discipline you are willing to impose on the factory will, in fact, decide the kinds of flexibilities you will be able to enjoy. Being permissive will only cause the factory to gum up, to lose the flexibility that permissiveness seeks in the first place.

SLOTH

Finally, there is sloth, the only deadly sin carried over from medieval times to current manufacturing. Sloth, of course, is a sin because in the factory as elsewhere, "Cleanliness is next to godliness." Cleanliness promotes visibility of what is done in the factory. It promotes the workforce's ownership of the process. People are much more willing to go the final mile if, in fact, the factory is a pleasant and neat place to work in.

Sloth, however, is a sin for more than its housekeeping implications. Sloth is a sin because it can mean failure to get the workforce involved and participating. We have come to learn how effective workforce participation can be in improving competitiveness and productivity. Slothful management does not tap into the innovation and wisdom that exists in the workforce. A company's unwillingness to work hard at participation tempts fate. Recognizing sloth as a sin emphasizes that we want the workplace clean but that we want management's hands dirty, dirty from continual visits to the factory to pat the backs of the workforce and to gather notes and suggestions from them.

THE MATCH-UP

The seven sins of manufacturing—inconsistency, complication, waste, meandering, impatience, permissiveness, and sloth—match up fairly well against the medieval admonitions. Pursuit of excellent manufacturing requires a look at the practical implications of the seven sins. A manufacturing strategy derived from cross-functional examination of the company's strengths and weaknesses can only be reinforced by attending to day-to-day operating concerns.

APPENDIX:
THE MANUFACTURING AUDIT

Operations choices can be segregated into three broad categories: (1) technology and facilities, (2) operating policies, and (3) operations organization. Let us review these categories and choices in turn.

Technology and Facilities

These choices frequently involve large capital expenditures and long periods of time. These are the big decisions that do much to define the type of process employed.

   1. *Nature of the process flow*: Is the flow of product through the plant characterized as rigid, with every product treated in the same way? At the other extreme, is the flow a jumbled one, with products routed in many different ways through the factory? Or, does the flow of product through the process fall somewhere in between?

Are segments of the process tightly linked to one another, or are the connections between process segments loose? How quickly can materials flow through the process?

Is the process a "pure" one or is it a hybrid?

2. *Vertical integration*: How much of a product's value is a direct result of factory operations? Could or should production involve more or less integration backward toward raw materials or more or less integration forward toward customers?

3. *Type of equipment*: Is the equipment used general purpose in design, or special purpose? Can any special-purpose, seemingly inflexible equipment be linked together in innovative ways to yield production systems that are themselves more flexible than the equipment they comprise?

Is the equipment meant for high speeds and long runs? How flexible is it for changeovers in products or models and how quickly and easily can such changes be accomplished?

What possibilities exist for linking one piece of equipment with another for balanced and quicker throughput?

Is the equipment operator-controlled or is it automatically or computer controlled? Is its performance monitored by operators or by a computer?

Can the equipment be speeded up or slowed down to match production needs?

Does the equipment demand substantial nonoperation support (e.g., maintenance, repair, software, setup, tooling)?

Can the company build or modify its own machines? Does it have close ties to equipment manufacturers?

4. *Degree of capital or labor intensity*: To what degree has the equipment or technology permitted labor value to be driven out of the product or service? How important could reduced labor value in the product be to costs, yields, or sustained levels of "production to specifications" (good quality)?

5. *Attitude toward the process technology*: To what extent does the company pioneer advances in process technology? Is the company a leader in that regard, or a follower? How closely does it track other process improvements?

How committed is the company to research and development? What mix is sought between the two?

How close is the alliance between manufacturing and engineering? To what extent are efforts made to design products that are easily manufactured? How are design and manufacturing teams organized for new-product introductions?

Are engineering change orders numerous? To what can they be attributed? How disruptive to the process flow are they?

What investments are made in manufacturing engineering and industrial engineering relative to product-design engineering?

6. *Attitude toward capacity utilization*: How close to capacity (defined as best as possible) does the company desire to operate? How willing is the company to stock out of a product because of too-tight capacity?

Does capacity generally come in significant chunks? What can be done to keep capacity well-balanced among segments of the process?

7. *Plant size*: How big does the company permit any one plant to grow? To what extent are either economies or diseconomies of scale present?

8. *Plant charters and locations*: Does it make sense to assign different product lines, product processes, or geographic markets to particular plants? How do the plant locations chosen mesh together into a multiplant operation?

Operating Policies

Once the process technology and facilities have been selected, management must still decide how the process technology is to be used. Three broad segments of such operating policies present themselves: loading the factory, controlling the movement of goods through it, and distributing the goods.

Loading the Factory

1. *Forecasting*: To what degree is the plant's output mix known with certainty before raw materials must be gathered or equipment or manpower assigned?

To what extent must forecasts be relied on to determine which raw materials ought to be ordered and which equipment or worker capacity reserved, and how much of each? How reliable have past forecasts been?

Does or should manufacturing second-guess marketing's demand forecasts? What techniques do best at forecasting requirements? Is there a long product pipeline that risks pipeline momentum problems and heightens forecasting needs?

2. *Purchasing*: Given the decision on the plant's degree of vertical integration, what does the plant make for itself and what is purchased from outside suppliers? How are such suppliers chosen? What kinds of contracts (e.g., long-term versus spot) are sought?

Is purchasing formally integrated with forecasting or order taking, and how much visibility is granted suppliers about the company's expected future needs? Are orders on suppliers made through an MRP system or other informal means, or are formal purchase requisitions required?

How important is supplier quality as against price? How and how well is supplier quality monitored?

3. *Supply logistics*: How frequently, from where, and by what mode of transportation do raw materials arrive at the plant? How sensitive are costs to changes in these factors?

How readily can materials be received, inspected, and stored? Is vendor quality good enough to dispense with incoming inspection? Is vendor delivery reliable enough and frequent enough to be able to feed the plant's production needs without large raw materials inventories?

How is materials handling accomplished within the plant? How much automatic, how much manual? What controls are in place?

4. *Raw materials inventory system*: How much inventory of raw materials is held? What system is used (e.g., material requirements planning (MRP), reorder point, periodic reorder)? How does the inventory level vary with demand, price discounts, supplier lead-time changes, supply uncertainties, or other factors? What triggers the replenishment of the raw materials inventory?

How are materials controlled in the stockroom? Are records accurate enough to be relied on without ceasing production for physical inventories? All scrap reported?

Are materials placed into "kits," or are they fed directly to the shop or line?

5. *Production planning*: Are goods manufactured to customer order, to forecasts, or are they manufactured to stock finished-goods inventory? Is inventory permitted to be built in advance to cover peak-period demands and thus smooth out production, or does production try to "chase" demand, with little or no build-up of inventories?

How are manpower needs planned for? How are model or product line changes planned for? How far in advance can changes be routinely made to the general production plan? How disruptive to the process is any expediting or rescheduling? Are routine allowances made for such situations?

Controlling Movement through the Factory

1. *Production scheduling and inventory control*: What triggers the specific production of goods: orders, forecasts, reference to a finished goods inventory? How do factors like the pattern of demand and product costs influence any trigger level of finished-goods inventory? How are specific departments, lines, workcenters, and the like scheduled, and what factors of the process, the pattern of demand, or product variations and costs affect the schedule?

What determines specific priorities in the jobs awaiting work in a department? How much expediting is permitted? How much rescheduling?

What or who determines the specific priorities of work: an MRP system, foreman's discretion, specific rules, simulation results?

Does instituting an MRP system make sense for the plant? How much fluctuation or stability exists for bills of materials, vendor reliability on delivery, production cycle times? What levels of accuracy and integrity exist for inventory counts and records, for production or rework counts and records for scrap? What kinds of exception reports and follow-through stand to be most helpful?

2. *Pacing of production*: Is the pace of production determined by machine setting, worker effort, management pressure or discretion, or some combination? How readily can the pace be altered?

3. *Production control*: How much information, and of what type, flows within the production process, both from management to the workforce and from the workforce to management? How easily can product variations, engineering changes, product-mix changes, or production-volume changes be transmitted to the workforce? How soon can management react to machine breakdowns, parts shortages, or other disruptions of the normal flow of product and process on the plant floor?

How are machines, workers, materials, and orders monitored? What "early warning signals" are looked to? What remedies are routinely investigated?

4. *Quality control*: Does everyone in the organization truly believe that quality (performance to specification) is his job, and not simply the function of a quality-control department? What mechanisms and cooperative forums exist to ensure that "work is done right the first time?" How closely linked are design engineering, manufacturing or industrial engineering, quality control, workforce training and supervision, maintenance, and production scheduling and control?

How is quality checked? How many checks are made at different steps in the process? How much authority is given to quality-control personnel?

5. *Workforce policies*: What are the skill levels required of various jobs throughout the process? How are they trained for? Is the work content of jobs broad or narrow? Is cross-training of workers desirable? How do workers advance in the factory (e.g., job classification, different jobs, changes of shift, promotion to management)?

How and how much are workers paid? Are any incentives, wage or otherwise, built into the process? How are workers' achievements and ideas encouraged and recognized?

How does management feel about unionization, and what actions concerning unionizations does it take?

What is the age and sex composition of the workforce?

What opportunities exist for job enlargement, job enrichment? Is there a place for quality-of-worklife projects? How else can the workforce be encouraged to participate in the management of the operation?

Distribution

1. *Distribution and Logistics*: What are the channels of distribution and how are they filled? What are the trade-offs of service versus inventory costs?

What are the benefits and costs of various geographical patterns of warehousing and distribution? What modes of transportation make sense and how should they be managed?

Operations Organization

1. *Operations control:* Are the major operating decisions retained centrally, or dispersed to individual plant units? What kinds of decisions rest primarily with the plant?

How is the plant evaluated, and what biases might the education method introduce?

2. *Talent:* Where within the organization are the best people put? What talents are most prized for the smooth and continued successful operation of the process?

NOTES

1. This term is credited to Wickham Skinner.
2. See Robin Cooper and Robert Kaplan, "Measure Costs Right: Make the Right Decisions." *Harvard Business Review,* September/October 1988, pp. 96–103.
3. See Roger Schmenner, "Escaping the Black Holes of Cost Accounting." *Business Horizons,* January/February 1988, pp. 66–72. A worked example is included in this article.
4. See Roger Schmenner, "The Merit of Making Things Fast." *Sloan Management Review,* Fall 1988, pp. 11–17.

CHAPTER 16

INTEGRATING BUSINESS AND MANUFACTURING STRATEGY

Frank S. Leonard

Editor's note

Why have so few companies failed to develop a successful manufacturing strategy? This chapter explores the impact on manufacturing operations of other functional activities, from sales through distribution. The author's case examples illustrate how the process can be rendered more effective by concentrating on making it interfunctional.

It has been over two decades since the idea of a manufacturing strategy was first widely circulated. In those 20 years, despite some dramatic successes, the stark reality is that most attempts to develop a manufacturing strategy—to reconfigure manufacturing to improve performance significantly—have fallen woefully short of expectations. Although there are a multitude of reasons for the lack of success, probably the greatest single mistake made by companies in trying to develop a manufacturing strategy is to consider the process as just a "manufacturing problem" or a "manufacturing responsibility." This error and its subsequent ramifications are responsible for a majority of derailed or impotent manufacturing strategy efforts. It is also responsible for the high degree of frustration and confusion manufacturing managers have reported in this process.

THE STRATEGY FORMATION PROCESS

In the broadest and deepest sense, the development of a competitive manufacturing strategy is a business unit problem, not a functional problem. A business unit is an organization capable in its entirety of all functions required to forecast, market, finance, design and engineer, produce, and distribute product. A functional approach segments the product design and delivery process into separate functional areas, such as finance or engineering. Each area may be moving in opposite directions from the corporate goals or from each other.

An interfunctional approach would therefore attempt to match up, or align, parallel objectives between various interest groups. A company interested in gaining a competitive advantage as a high-tech producer dependent on fast new-product introductions might concentrate its strategy efforts on alignment of engineering and production goals. An engineering department in a company handling strategy formation intrafunctionally might, for example, pursue numerous design changes in the form of engineering change orders (ECOs) at all costs. Excessive use of ECOs throughout the product life cycle adds to production costs and delivery problems.

When viewed as an interfunctional process, the manufacturing strategy process is a valuable corporate tool to explore critical interdependencies in a business unit, to align them, and to build them into formidable competitive advantages.

The alignment of major elements within manufacturing proper toward strategic requirements is a critical step in the manufacturing-strategy process. However, forces outside the manufacturing function are what often determine the effectiveness of this alignment—the ability of manufacturing to significantly increase its strategic contribution. Manufacturing's ability to change, to align itself strategically, is severely, though often subtly, hindered by these external forces. They are forces, such as marketing pressure or capital-investment policies, not easily understood by manufacturing without a great deal of insight and effort.

Manufacturing Strategy as the Integrator

When seen as an interfunctional instead of an intrafunctional process, manufacturing strategy's real power is its ability to integrate the rest of the business with manufacturing. It can, for instance, ensure that the R&D strategy, marketing programs, procurement policies, technology imperatives, and management development process are aligned with manufacturing capabilities. The influence of other parts of the organization on manufacturing, from the obvious (the sales forecast) to the more subtle (the capital-appropriation procedure) to the almost invisible (accounting practices) is pervasive. World-class manufacturing firms have already learned the necessity for understanding the nuances of this interfunctional alignment. World-class manufacturing firms are using the manufacturing strategy process as an important vehicle to obtain synergistic benefits from the coordination of various parts of the business unit, as well as to strategically position the major elements within manufacturing decision areas.

THE INDEPENDENT-FUNCTION ARGUMENT

There are, however, executives and consultants who believe in a contrary argument; let me call this the "independent-function argument." This argument states that the manufacturing strategy concerns only manufacturing decisions; it can and should be formulated by the manufacturing function operating independently. This argument is based on the idea that the various functions and departments in a business unit are primarily independent of one another, that their spheres of decision making (and hence their capabilities) are relatively free from decisions in other functional areas. Pursuing this argument further, if the other functions (or the business unit as a whole) state their strategies and policies clearly, then the process of developing a manufacturing strategy can be accomplished entirely within manufacturing by using these strategies and policies as decision parameters and constraints. To the people who believe in the independent-function argument, it is the strategic planning process that should and does integrate the business unit. In other words, strategy drives manufacturing—a kind of top-down approach.

The independent-function idea is based on several tenuous if not fallacious assumptions about the reality of modern corporations. First, it assumes that the rest of the company's policies and functional strategies are correct and that manufacturing needs to fit into them. An interesting psychological corollary to this is that if manufacturing is the "misfit," then somehow manufacturing is "wrong."

Second, the independent-function argument assumes that manufacturing capabilities cannot be influential, much less critical, in defining and achieving market positions or in defining other functions' policies. As a counterexample to this assumption, if manufacturing is capable of delivering in two rather than four weeks, the location and stocking of distribution warehouses would be substantially different—fewer, perhaps, more spread out, and smaller.

Third, the independent-function argument assumes that the policies and strategies of other functions are compatible with one another and with manufacturing, that they have no internal contradictions that would make it impossible for manufacturing to develop a consistent strategy to fit them all. New-product introductions often display conflicting policies between R&D and marketing, leaving manufacturing without consistent time schedules.

Finally, it assumes that the independencies among functions are more critical to competitive position than the interdependencies.

The above assumptions may have fit the corporation and organizational structure of 1950, but they do not reflect the reality and complexity of today's corporations, technologies, and global competition, in which decisions affect all functional areas.

In the fullest sense, the manufacturing-strategy process is a process of mutual adaptation among the various parts of a business unit. It is a process of mutually adapting manufacturing capabilities, realized or unrealized, with, for instance, engineering, marketing, sales, labor relations, management development, and strategic planning policies and procedures. It is a process of discovering intrafunctional and interfunctional contradictions and synergies that are normally left buried due to the nature of most corporate strategic planning processes. To be successful, the manufacturing-strategy process must be a stimulus for examining the critical alignment of major functional strategies and business policies.

FORCES WORKING AGAINST THE PROCESS

There are many forces, however, operating to isolate the manufacturing-strategy process as just a manufacturing responsibility. First, manufacturing has traditionally physically isolated itself from the rest of the organization; plants are often distant from many other corporate offices.

Second, there has been little attempt by top management to focus on cross-integrating functional strategies. They assume that the strategic planning process will accomplish this integration. Most chief operating officers, unfortunately, are not from manufacturing, and they tend to want someone else to manage the details. The reality of most strategic planning is that it is a bottom-up collation of functional plans and budgets that rarely, if ever, fine tunes interfunctional alignment below the top policy levels.

Third, a common language, what Romeyn Everdell calls a Rosetta stone, is often lacking to translate across functional lines. Each function has its own specialist jargon and perspective, which precludes a good understanding of other functional requirements and problems. Sales speaks in sales dollars; manufacturing, in cost of goods sold or another unit measure; engineering may not be concerned with dollars at all; and cost accounting may concentrate on variances!

Fourth, the inevitable turf issue raises its ugly head: Functional managers like to keep control over their areas. They feel they are the experts in their areas and have little to gain from discussions with others.

Finally, the manufacturing strategy process is time consuming, requiring a lot of patience and mutual exploration of conceptually cloudy issues without rigid methodologies. The process may take managers off site for entire days, sometimes monthly, but at least quarterly. It is not a tidy, fill-in-the-blank process. This is hardly the type of activity busy executives find time for or are expected to do.[*]

Although many managers might prefer it, manufacturing can no longer operate as an independent member of a business unit. Over the last two decades, the close interdependencies necessary among all functions have been increasing dramatically. Far Eastern and European global players, such as Toyota and BMW, Sony and Phillips, have made that quite clear. Major decisions made in engineering have a drastic impact on the ability of manufacturing to design new processes. Decisions made about sales plans have a large impact on manufacturing flexibility. Manufacturing decisions on plant location have an important effect on parts availability of service centers. The efforts over the last five years on company-wide quality programs and multifunctional team approaches to product design are important indicators of the growing awareness of these critical interdependencies. Some theorists have called the need for planning to integrate all functions a "holistic approach."

---

[*]Editor's Note: The information gathering and manipulation work is not easily computerized without good planning tools. See Chapter 13 for an examination of a very simple planning process.

Making the Connection: Interdependencies outside Manufacturing

There are many interdependencies outside of manufacturing—in engineering, marketing, sales, finance, human resource management (HRM), and accounting—that influence and can be influenced by the manufacturing strategy. Sometimes they are company-specific in their impact on manufacturing. There are, however, some critical ones that, time and time again across many industries, inhibit manufacturing's ability to make significant strategic strides. A brief description and comment will help to illuminate the nature of these interdependencies. With any of these influences there are two key questions to keep in mind:

1. How does this interdependency shape competitive manufacturing capabilities?
2. How does this interdependency prevent manufacturing from doing something that might be turned into a competitive advantage?

Engineering

The most common processes that affect manufacturing from the engineering or R&D area are the new-product development process and the engineering change order (ECO) process. The first determines the design of a new product, and the second concerns the change of an existing product. Although many companies are doing a better job of coordinating manufacturing concerns in the new-product design process, there is still a tendency to keep most decisions in the hands of technical people. Manufacturing fills a "veto" role, getting a change, as one manager put it, "only when I scream loud enough and long enough." The impact on manufacturing of these design decisions influences critical issues such as process design, process improvements, scheduling, plant configurations, plant layout, and manufacturing organization and staffing.

A truly creative exchange exists in the new-product design process when both engineering and manufacturing assume all design parameters are tentative and open to negotiation, including materials, shapes, technologies, and functions.[*]

---

[*]Editor's Note: Both Black and Decker and Digital Equipment Corporation use product-development teams that, from the beginning, include all functional areas and try to speed the cycle.

Manufacturing is often constrained most by existing ECO processes. Manufacturing managers frequently have no say in the number or frequency of ECOs. An indisputable (sometimes inscrutable!) logic seems to prevail when it comes to ECOs. Many companies have no priority weighting for ECOs. ECOs can be generated by both internal and external sources with no decision screens to evaluate their effectiveness. Yet the ECO process is the major interruption preventing manufacturing from improving its present processes. The effect of ECOs on the stability and improvement of existing processes, and on manufacturing resources, is dramatic but frequently overlooked. One company had so many ECOs on a certain product that required tooling changes that, by the time the new tooling arrived, it was already obsolete. There was over $50,000 of tooling in inventory that had never been used.

A more subtle influence often overlooked is the relationship between the new-product design process and the ECO process. It is difficult to change one without changing the other; they are symbiotically self-perpetuating.

Marketing

Marketing, "the voice of the customers," has a pervasive influence on manufacturing that is too often assumed to be positive in its present form. Frequently affecting manufacturing are warranty and service programs that marketing institutes and changes. Drastic changes in service policies and warranty policies are often instituted by marketing without any discussion of the impact on manufacturing. The design of product-testing equipment, the levels and types of inventories, the relationships with vendors, and the scheduling difficulties are all downstream influences that manufacturing too often has to swallow. One marketing department of a specialty chemical manufacturer instituted a marketing program based on assuring the customers a higher level of quality (fewer impurities) although the existing manufacturing process was technically unable to achieve the advertised level of product purity.

Market segmentation. A more subtle impact is the language and concepts employed in the marketing strategy. All too often a segmentation by product or customer is effected, included in the marketing strategy, and submitted to top management. The assumption by top management is that the segmentation used for the marketing strategy is adequate, even correct, for manufacturing strategy. This can be a major source of confusion in developing a manufacturing strategy.

One basic chemical company had segmented its markets into a pulp and paper market and the water treatment market. The marketing manager and the general manager then asked manufacturing to come up with a manufacturing strategy for these two markets. On deeper examination it was obvious that the two segments that would be relevant for building a competitive manufacturing strategy were markets based on manufacturing requirements. In this instance there were two segments, but not the pulp-and-paper and water-treatment segments: The two segments for manufacturing purposes were the markets where delivery was a high priority and the market where cost was a high priority. It turned out that these two segments cut across both the pulp-and-paper and water-treatment segments.

Sales

The most frequently mentioned problem affecting a manufacturing strategy from the sales area is the forecast. The predictability, variation, and time lag of the forecast affects manufacturing schedules, line changeovers, process designs, process improvements, equipment purchases, plant expansions or contractions, hiring and training of the workforce, and plant configuration. Once again, like marketing, it is too often assumed in the company (by both manufacturing and sales) that manufacturing must do whatever is required to fill customer orders, regardless of the forecast. This is frequently destructive to the consistency necessary for a manufacturing strategy to evolve.

In one company it took the general manager to intervene to prevent the sales manager from accelerating a plant start-up because of a "sudden influx" of orders. The result of accelerating the start-up (long-run reliability and warranty problems) would have killed the product two or three years down the road. Yet it was common practice in that company for the sales manager to determine the speed of new-product introductions according to short-run sales demand rather than through the evolution of a process/product combination.

Finance and Accounting

The two primary influences from finance and accounting are the capital-appropriation process and the cost-allocation scheme. The capital-appropriation process, including the rationale, the paperwork, the hurdle rates, the review procedures, and the inevitable politics surrounding it, has both an obvious and a subtle effect on manufacturing. At the obvious level, the quantitative payback expected (often informally communicated or implicitly understood) determines exactly what projects and what parts of what projects will be considered. This affects plant expansions, process improvements, capacity changes, and most recently, systems development and applications. It forces manufacturing to discard any long-term effect that cannot be quantified or proven.

It further forces manufacturing to submit only portions of a total project in order to get the required payback. I call this phenomenon "chopped projects," and its result is that the infrastructural supports for an investment are not there when needed.

In order to hit the hurdle rate, one company decided to submit a project only with the assumption that the process would run five years without a major shutdown and overhaul rate. Not only did this give them more productive time but they did not need to include all of the maintenance costs for the three-week shutdowns. When, however, the plant had to go through a shutdown in the second year, not only did it take six weeks instead of three, it cost twice what it would have had there been proper maintenance people and spare parts. The plant start-up after the shutdown took twice as long as budgeted because the present plant management had not gone through a start-up. As a result, short-term operating results were disastrous, not planned for in the project payback estimate of two years earlier.

The accounting basis for cost allocation and subsequent product-cost determination, with its influence on management decision making, is currently under much scrutiny by the accounting community. These allocation schemes influence manufacturing decisions about staffing levels, process improvements, product-line phase-outs, benefit and compensation for the workforce, and plant configurations—without accurately reflecting true product costs. Financial and accounting procedures are often dictated by corporate-level staff groups far from the manufacturing floor who have little knowledge or concern with the inevitable impact on manufacturing decisions.

Usually connected with the accounting system is the system for measuring manufacturing performance. The aphorism, "You get what you measure," is unfortunately too true. You can be sure a plant justified, built, and measured on 90 percent utilization will get a 90 percent utilization—at the expense of quality, delivery, and flexibility.

Human Resources Management (HRM)

The influence on manufacturing strategy of HRM groups is more sub-tle but just as pervasive. Benefit programs, management development, training and education programs, job placement, and manpower plan-ning (to name a few) developed by human resource departments have an important effect on just how manufacturing strategy can use both the workforce and management people to achieve competitive advantage. The human resource policies and procedures influence how manufactur-ing recruits and hires people and how they promote or rotate them. They also influence the level, type, and quality of training and educational programs available, job definitions, and, where appropriate, labor union work rules. Often locked up in inviolable policies that defy changing, manufacturing is left to use its people in highly restricted and unsatis-factory ways.

In one very large multiplant complex with multiple business units, the company agreed to cover all labor relations by a single contract with standardized work rules. This was advantageous to both the union leadership and the labor relations department.

On close examination of workforce behavior as it related to compet-itive performance, it was quite apparent that the various manufacturing strategies of the businesses in the complex were drastically disparate: One needed a highly skilled, very flexible workforce with a lot of tech-nical training; a second needed a very low-skilled workforce with only minimal training at time of hiring; a third needed a workforce that com-bined the operators and maintenance people into discrete work groups with as little turnover as possible. Yet the labor policies of this complex did none of these very well. They were, unfortunately, a compromise of political astuteness and competitive mediocrity.

It is a sad fact that 50 years of insight into human relations within business has had little effect getting the traditional, hierarchical, and narrow HRM approaches to use people creatively.

Strategic Planning

Finally, the corporate planning departments influence how effective a manufacturing strategy effort can be. Through their procedures and pro-cesses, they can either stifle or enhance the manufacturing strategy process. The planning process can either include the language and con-cepts of manufacturing strategy or exclude them. Most importantly, because most companies can't afford the time to go through the planning process twice, once officially for corporate bureaucracy and once for their own fine tuning, it is the depth and quality of the strategic planning process, overseen by corporate analysis, that determines the quality of the entire process. Is it just a bottom-up "roll-up" of budgets and projects, or a truly interfunctional process of exploring critical interdependencies among various parts of the business? If the process is a roll-up, the chances are very high that the manufacturing-strategy process, even if sanctioned at corporate level, will be an intrafunctional process with subsequent narrow results. It will be something put together from figures (profits, projections, expenses, etc.) compiled by individual departments on their own. Their first look at the big picture comes when the spreadsheet produces the roll-up.

If, however, the process is an interfunctional one, there is a good chance that the manufacturing strategy will make use of interfunctional synergies.

THE GOOD NEWS

A manufacturer of compressors decided to try to manage manufacturing strategically and found the road to significant change was a long and successful one. Over the course of five years, not only did they restructure the entire architecture of manufacturing, but they also succeeded in discovering and then changing critical interdependencies outside of manufacturing. They realized that they could not reconfigure manufacturing without also reconfiguring interfunctional influences.

In five years they changed the entire *sales forecasting procedure*: Any changes in schedule made by sales within 30 days of production caused a penalty payment to be made to manufacturing from the sales department's budget, a kind of "forecast-change variance."

The *process of introducing new products* was entirely revamped: All new products are developed by a joint team, headed by either a manufacturing or quality person. The team does not report to engineering or manufacturing; it reports to the general manager of the business.

The *criteria* on which manufacturing managers were evaluated and promoted were changed from "hitting budget" to increased competitive performance along five specific dimensions.

The *capital-appropriation review process* was altered from an entirely quantitative process to an equal mix of qualitative and quantitative issues.

The *measurement of performance* for the plants was elaborated: Each plant had specific, unique evaluation criteria depending on its strategic focus.

Finally, the *strategic planning process* was expanded. It now includes a thorough review of each department's goals and concerns long before the plans are rolled up to top management. As a result, manufacturing has found that it is actually helping set strategic direction.

These changes have taken almost five years first to understand and then to begin to implement. They required direct involvement of the general manager. They cost some managers their jobs, both in and out of manufacturing, but they resulted in catapulting the company into being a world-class manufacturer competing with the best Asian and European companies.

THE BAD NEWS

The shift of manufacturing from more traditional stances and responses to world-class competition is not smooth, easy, or short. Experience shows that far-reaching and significant changes must be accomplished: first, in manufacturing; then, in how the corporation manages manufacturing; finally, in how manufacturing influences the corporation. The external influences on manufacturing detailed in this chapter can be considered either as iron-clad constraints not open for discussion and alteration, or they can be seen as simply arbitrary and provisional corporate influences that can be adjusted appropriately to allow manufacturing to realize its strategic capabilities fully.

The process of understanding and managing the interfunctional aspects of strategy is not tidy; it is very real. Even if you come up with a good interfunctional manufacturing strategy—a piece of paper and a collection of ideas—you are still not home free. You have to make those ideas come alive through your organization; the ideas must be turned into forceful action. There are no simple recipes or standard paths in this process. But if you want to boost the strategic capability of manufacturing (and there are many advocates who say you probably have no choice in the matter), you will have to address concerns and issues external to manufacturing.

Pursuing a manufacturing strategy means far more than putting together a three-ring binder filled with analysis; it means challenging the very standards by which your organization evaluates and judges manufacturing performance. It means making the standard of manufacturing excellence your strategic agenda. It means relentlessly altering anything in the corporation that hinders the accomplishment of that agenda. It means change and all that change implies: effort, time, and money for corporate achievement and human satisfaction. Now, that isn't entirely bad news, is it?

CHAPTER 17

COMPETING IN GOOD TIMES AND BAD

Ed Heard

Editor's note

What will it take for the U.S. to be competitive in the next decade? Being competitive and winning require addressing issues of quality, price, customer service, and especially responsiveness. Management decision tools (e.g., financial ratios) are not sufficient for deciding how to survive good or bad times. The author urges manufacturing management to address other critical factors—success models, capacity planning, and performance drivers.

When it comes to increasing competitiveness, our business and political leaders appear to be intellectually bankrupt. Shareholders, stock analysts, and the financial markets demand short-term results. But woe be unto the economy whose leaders begin to see the world through financial blinders. Market restrictions, currency exchange rates, and wage levels become kneejerk explanations for poor competitive performance. In the United States, the government serves as an all too convenient whipping boy, while top managers try to force the numbers to come out right through relentless rounds of divestitures, mergers, and acquisitions.

So what? Declining market share and low profitability are not problems to be corrected directly; they are symptoms to be investigated. Current quality, cost, and investment levels are results of numerous past decisions—not just those made since the last accounting period.

The real question is what guided those decisions. Does top management subscribe to an evolving but consistent cause-and-effect vision of what it takes to compete successfully in world markets? If it does, has it shared that vision with the company, community, and country widely and repeatedly?

The fact that North American manufacturing is under siege is not secret. First came the cost battle. The quality battle is still going on. Responsiveness is probably next. And after that? Variety, responsiveness, high quality, and low cost—all at the same time. This paper will first describe the nature of the responsiveness battle. It will then discuss what the individual manufacturing company has to do to win the responsiveness struggle and position itself for the next battle.

As markets grow ever more competitive, most companies have attempted to rise to the challenge. Some have succeeded; others have dropped by the wayside. In many markets there is little price and quality differentiation among those that are left. Low cost and high quality are becoming prerequisites to survival instead of competitive variables.

At the same time, markets are becoming more volatile and product life cycles are getting shorter. The new battle ground is responsiveness. As new markets develop (or as global markets open up), the company that can get there first has the advantage. Likewise, the company that makes the quickest response to true demand shifts has a real edge, as does the firm able to respond to specific customer requests on short lead times.

What does all this mean for the individual manufacturing company? It must:

- Shorten new-product introduction lead times.
- Shorten manufacturing-cycle times.
- Shorten sales and distribution lead times.
- Shorten white collar cycle times.

WHAT IS COMPETITIVENESS?

Competitiveness is a term frequently used but rarely defined. A dictionary isn't much help. Competitiveness is defined as "having the property of being competitive." In turn, *competitive* is apparently defined as "determined by competition." Finally, *competition* refers to "individuals, teams, organizations, etc., vying for the same awards, goals, markets, profits, etc."

Examined within the business context, where the word *competitiveness* appears to be most often used, those definitions are not very satisfactory: A company could be described as highly competitive while clearly on the path to bankruptcy.

Price, quality, and service have long been recognized as primary determinants of the relative strengths of competitors in specific markets. In turn, product cost, quality, and availability have long been seen primarily as manufacturing responsibilities. Past performance, however, indicates that individuals in the manufacturing community have differed greatly in their interpretations of and their approaches to those responsibilities.

At this point, it seems clear that for several decades low product cost was pursued much more rigorously than product quality and availability. Likewise, it seems clear now that the need for product availability came to be viewed one-dimensionally. "Deliver as promised" became the battle cry. Somewhere we lost sight of what the customer wants— *responsiveness and dependability*, not just dependability of product or delivery.

A whole generation of manufacturing professionals grew up thinking that all there is to customer service is on-time delivery. For many years, few manufacturers worried about the determinants of responsiveness.* Indeed, much of the energy of the last two decades was spent developing and implementing ever more elaborate planning and scheduling systems to cope with the increasing variability in the largely untended manufacturing environment.

Competitiveness is about winning. What is winning? In athletic contests, games of chance, horse races, or shoot-outs, the winner is obvious. It's the one still standing, the one who gets to the finish line first, or the one who accumulates the most points. What about league

---

*Editor's note: Indeed, we measure good customer service as percent of product shipped on time complete *to promise*. Few performance measurements rate excellent customer service as shipment on-time complete *as requested* (or to the customer's original request).

In addition to the fact that we often fail to capture the customer's original quantity and ship-date request at the time of order placement, if we do not retain that critical piece of data in the demand-history file used to produce new forecasts, we will project forward that same "unresponsive" ship schedule.

championships, play-offs, and the like? Winning in those contests depends on the rules governing them. If the loser of each individual event is automatically eliminated, the winner is the only entrant undefeated at the end. If every entrant participates in the same number of individual events, the winner may be the entrant with the highest winning percentage. The possibilities are endless!

There are major differences between competing in contests and competing in global markets. First and foremost, the rules are a lot clearer in sports. Second, clocks, calendars, or number of events play significant roles in sporting events. Almost without exception, such contests have defined endpoints. Likewise, league championships and play-offs are restricted to specific teams or numbers of participants. By contrast, the number of competitors in global contests is constantly changing; there are no limits on time or number of events, and the rules are not very clear. If in fact there is only one winner, the identity of that winner is not readily apparent.

The losers, however, are obvious. They are the enterprises forced to withdraw from the game. Sometimes their products won't sell. Sometimes their costs exceed their revenues. Sometimes departures are blamed on undercapitalization. More often, it is reported that all of these factors played a part in their defeat. Clearly, enterprises that are forced out of the game cannot be called competitive. Likewise, economies where such enterprises are the rule instead of the exception do not deserve that label.

What about enterprises that continue to be significant players in their respective global markets year in and year out, decade after decade? What about economies where such enterprises are common? Clearly, such enterprises and economies are highly competitive.

A working definition of global-market competitiveness is not far away. Competitiveness is about winning. Winning is about surviving and prospering through the good years and bad. Clearly, competitiveness in a global marketplace is the ability to continue to survive and prosper regardless of changing world conditions.

There are five cycles that determine competitiveness: the book/bill cycle, the purchase/produce cycle, manufacturing-cycle time, the design/develop cycle, and the spec/source cycle. When shortened, each of these cycles, which add up to what has been called *cumulative lead time*, can enhance a competitor's responsiveness, hence his chances of winning and retaining customers.

The Book/Bill Cycle

Most of the activities that make up the book/bill cycle are white collar activities. Customer requirements must be translated into actual sales-order and delivery promises. In turn, sales orders must be added to the open order file. Customer credit must be checked. Pricing must be verified. Even if the necessary products are available on the shelf, picking, packing, and shipping must be scheduled. Pick lists and shipping documents must be prepared. Ultimately, the order must be picked, packed, and shipped, and the customer must be invoiced.

These activities take time even when the necessary products are available on the shelf. But what if they are not? In some cases, inventorying finished goods may be economically impractical. In other cases, products may have to be assembled, configured, or fabricated to order. In the extreme case, products may even have to be engineered to order.

Regardless of the reason that orders cannot be filled from stock, the net effect is the same: The book/bill cycle consists of a larger number of activities, and the critical path through them is longer. Whether long or short, the book/bill cycle represents a discrepancy between when the customer would really like to have his order and when he can actually get it. Consequently, book/bill cycle time is a major competitive factor.

The Purchase/Produce Cycle

Even if sources have already been identified and long-term agreements with them established, actual releases (requests) must be initiated and transmitted to suppliers. Materials must be shipped from suppliers to the enterprise's facility, where they must be unloaded and received. Packing slips must be verified against releases; orders must be unpacked; counts and quality must be verified; and inventory transactions must be completed before materials are put away or taken directly to the line for use.

If materials are stored, some form of requisition must be prepared to trigger removal. Pick lists may have to be generated and storage locations identified. Material often has to be picked, counted, and issued, and inventory transactions usually have to be prepared before material leaves the storeroom and is transported to the places where it is used on the floor.

On the floor itself, materials must be processed, inspected, counted, and moved. Machines must be maintained, changed over, and cleaned. Parts must be counted and inventory transactions prepared. Tools, fixtures, and dies have to be located, retrieved, transported, and put away. Orders must be opened and closed. Typically, the finished product must be checked and approved for shipment before it can be picked, packed, and loaded for shipment.

Manufacturing-Cycle Time

Material must move from the supplier through the factory before it can become part of a shipment to a customer. After material arrives at the receiving dock, many internal physical and information-processing activities must take place before that material can complete its journey. The critical path through all internal activities is usually referred to as the *manufacturing-cycle time*.

Manufacturing-cycle time is critically important to enterprise competitiveness. To be viable in the long term, the gross margin for an enterprise's products must be large enough to provide an adequate return on assets. Inventory is one of those assets. The longer the manufacturing-cycle time, the greater the amount of inventory. In turn, the greater the inventory, the larger the gross margin requirements are. Ultimately, the larger the gross margin requirements, the higher and less competitive the necessary price becomes.

For enterprises that must build in whole or in part to order, the purchase/produce cycle packs a double punch: Not only do longer manufacturing-cycle times, the major portion of the purchase/produce cycle, mean less competitive prices; they also mean less competitive response times. Why? Remember that any portion of the purchase/produce cycle that cannot be completed until the customer order is in hand is a part of the book/bill cycle, a major competitive factor.

Design/Develop Cycle

Some manufacturing enterprises engineer to order; they have no stock products as such. Still others introduce new products periodically for strategic reasons. The design/develop cycle is a major competitive factor in both cases. In the engineer-to-order case, the design/develop cycle is a major contributor to the book/bill cycle.

The potentially negative impact of longer design/develop cycles in new-product introduction is not well understood. When the nature of market demand cannot be predicted very far in advance, the inevitable result of long design/develop cycles is missed market opportunities. But when the nature of market demand can be predicted, why not just start earlier?

Regardless of the reasons for longer design/develop cycles, larger total expenses are generally the result. What is not so clear is that longer design/develop cycles also require more borrowing or diversion of profits to support the necessarily longer periods of negative cash flow. As a result, long design/develop cycles pose a triple threat to competitiveness: They limit responsiveness, raise expenses, and increase capital requirements.

What takes so long? Just a few incidental processing activities: market analysis, marketing product definition, technology analysis, engineering product definition, product design, prototype development, functional verification, engineering model development and evaluation, preproduction unit development, manufacturing performance evaluation, field tests, process planning, process verification.

Spec/Source Cycle

At some point, specifications for the materials used to produce new products must be developed. Fit, function, performance, quality, delivery, and price requirements must be established. Make or buy decisions must be made. Potential suppliers of purchased items must be identified and preliminary contacts initiated. Visitations must be arranged and completed. Suppliers must be evaluated and selection decisions made. Specific issues must be negotiated and long-term agreements worked out.

The spec/source activities take time and cost money. Like the design/develop cycle, the spec/source cycle is often part of the book/bill cycle for engineer-to-order products. Likewise, it is also part of the new-product introduction cycle. Long spec/source cycles contribute negatively to an enterprise's ability to compete just as surely and in the same ways that long design/develop cycles do.

Cycles in Typical Enterprises

The relative importance of the four cycles (five including manufacturing) is a function of their market environments. In the "normal" situation that most writers assume, the product line is described as relatively stable with periodic updates. Products are typically described as requiring multi-level fabrication and assembly. Likewise, manufacturing-cycle times are generally assumed to be longer than the response times demanded by the market.

In such environments, forecasts are used to drive long cycle-time material acquisition, fabrication, and assembly. When an order is received, the appropriate in-process materials are located, earmarked, and (ultimately) expedited to fill the order. Although some orders may be fillable from inventory, writers generally assume that most are not. The usual rationale is that inventory stocking is too expensive, that certain product features are order specific, or that demand cannot be predicted.

The importance of all four of the major cycles is clear in such enterprises. Since no one cycle dominates the others, the importance of integration and coordination of the functions is evident. Over time, the management of such enterprises has tried a number of "add on" approaches to filling that need (meetings, task forces, management by objective, matrix management, MRPII, etc). *The danger is that management can become so preoccupied with coping better that improvement efforts are left on the back burner perpetually.*

Cycles in Repetitive Enterprises

Such environments tend to be characterized by high-volume production and distribution of highly stable product lines (e.g., power mowers or aspirin). Model changes occur infrequently and at regularly scheduled intervals. Since these are commodity or near-commodity products, market conditions dictate off-the-shelf service. Forecasts, stocking policies, and current inventory levels are used to adjust production levels as necessary.

The stability and the high volume of the product line profoundly influence the way such enterprises evolve and operate. The need for low-cost production and a continuous supply of the right materials is obvious to everyone in the enterprise. It is no surprise that successful enterprises of this type develop a lot of muscle in purchasing and manufacturing, the dominant contributors to the purchase/produce cycle.

The flipside of the coin is that such enterprises tend to have relatively underdeveloped muscles in the parts of the organization concerned with the design/develop cycle. The spec/source and book/bill cycles also tend to be underdeveloped, but to a lesser extent. Management preoccupation with the purchase/produce cycle may well result in missed opportunities to improve responsiveness, cost, and quality in the other cycles. New-product introductions tend to take longer and longer and get costlier and costlier in such enterprises.

Cycles in Engineer-to-Order Enterprises

In the extreme engineer-to-order enterprise, almost all products are custom designed, and very few parts are interchangeable among them. Customer requirements must be clearly defined, conceptual product designs created, prototypes built and tested, and so on. Manufacturing is the last in a long series of activities that must be planned, scheduled, and coordinated. In this case, the book/bill cycle is all inclusive: Booking an order is the first thing that happens, and invoicing the customer is the last.

In such enterprises, the design/develop cycle typically overwhelms the purchase/produce cycle and, to a lesser extent, dominates the spec/source cycle. Here, too, management tends to become preoccupied with integrating and coordinating, but the focus is primarily on the activities that compose the design/develop cycle. Both the individual activities that make up a project, and resource commitments to competing projects must be planned, scheduled, and controlled.

In these enterprises, the design/develop cycle's muscles tend to be highly developed. Purchase/produce muscles seem almost atrophied by comparison. These enterprises have special problems, due to the disproportionately large percentages of white collar workers and the nature of the work they do. In general, techniques for managing white collar nonrepetitive work are not well-developed or well-known. The tendency in such enterprises is for the design/develop cycle to get longer and more costly; as a result designs become harder to produce.

Expanding Customer Requirements

Clearly, the historical trends in traditional manufacturing enterprises are inconsistent with marketplace trends. Not only is conflict between the two inevitable, it is already occurring in market after market around the world. High-quality and low-price products are becoming the price of market admission, and customers are demanding quicker response.

Industrial customers and the consuming public are becoming more discriminating. They are beginning to think total cost, not just product cost. Low prices accompanied by unwieldy order-entry and customer-service processes that raise the effective cost of doing business are not very impressive. Correct invoices, readable owner's manuals, and the like become the new parameters of quality once product excellence is ensured. Even the definition of responsiveness is changing.

Customers are beginning to notice employee attitudes, how long it takes to get questions answered, and how many times the phone rings before someone answers. In short, the customer is beginning to measure cost, quality, and responsiveness in terms of all the interfaces between himself and the enterprise. Clearly, enterprises must adapt to the new reality and begin to think:

Low cost, high quality and short cycles in everything we do!

A TOP MANAGEMENT PERSPECTIVE
ON SUCCESS INDICATORS

While competitiveness is clearly a long-term concern, many argue that top management doesn't have the luxury of looking very far ahead. According to this school of thought, chief executives live in fear of hostile takeovers. The recent wave of mergers and acquisitions lends support to this contention.

Top management typically understands very well that it must keep share price up, and that growth and return on equity (ROE) are critical to that end (see Figure 17–1). Beyond that, history seems to suggest a woefully inadequate understanding of the cause-and-effect relationships between top management actions and those numbers.

In today's world, there are few geopolitical obstacles to capital movement. Just as water seeks the lowest level, capital flows to opportunities that appear to offer the highest potential returns. Pity the top management team whose stock value drops below the current value of the assets of the enterprise.

How do companies protect themselves? Stock markets respond to growth potential, demonstrated ROE, and market share performance. Of these indices, ROE is the most easily accessible and understandable.

FIGURE 17-1
A Question of Survival

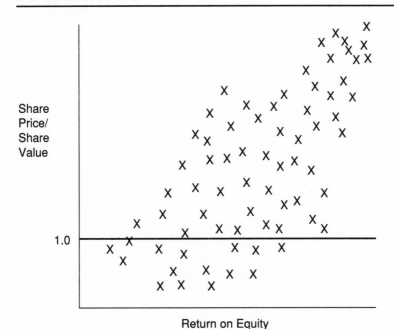

Return on Equity

Is it any wonder that top management becomes preoccupied with it?

$$\text{ROE} = \frac{\text{Profit}}{\text{Margin}} \times \frac{\text{Asset}}{\text{Turns}} \times \frac{\text{Financial}}{\text{Leverage}}$$

In fact, that's what many mergers and acquisitions are about. To appreciate that concept, ROE must be examined more closely.

$$\text{ROE} = \frac{\text{Profit}}{\text{Revenue}} \times \frac{\text{Revenue}}{\text{Assets}} \times \frac{\text{Assets}}{\text{Equity}}$$

The real objective of divestitures, mergers, acquisitions, and spin-offs is to create new enterprises with better looking ROE and market positions. In turn, a better ROE reduces some of the pressure on operating units to keep prices and profits artificially high. Top management often sees financial restructuring as not only a way of protecting its vested interests, but as a shortcut to increased competitiveness as well.

Restructuring focuses primarily on the financial-leverage portion of ROE and occurs infrequently. What about day-to-day operations? The profit-margin and asset turn components of ROE make up ROA (return on assets):

$$ROA = \frac{Profit}{Revenue} \times \frac{Revenue}{Assets}$$

Each element of the ROA equation has a significant number of components. How can the folks at the top possibly manage all of them (see Figure 17–2)?

It is simply not possible for top management to have a grip on every detail. Instead, responsibilities for the components of Figure 17–2 are parcelled out to several individuals. In turn, these individuals sort out their responsibilities for distribution among their subordinates, who delegate their responsibilities to still another layer of subordinates, and so forth. This practice results in the organizational pyramids so prevalent throughout industry.

Top management sees its job as making sure the numbers come out right. Goals are set for subordinates, performance is measured, variances are identified, righteous indignation is displayed, explanations are demanded, and justifications are submitted. Naturally, this process is repeated at the next level, and the next, and the next.

THE ROLE OF SUCCESS MODELS

Why is the process replicated so faithfully level by level? Because it is policy, either stated explicitly or widely believed to exist implicitly. It is not restricted to the financial control model described above, either. By repeated example, top management reveals its most closely held values and beliefs about what is required for a product-based enterprise to survive and prosper; that is, top management reveals its success model.

Top management doesn't have a monopoly on success models. At lower levels of the organizational pyramid, managers are responsible for *activities* as well as numbers. They select and operate processes, design and use procedures, configure and run plants, and so forth. Clearly, their choices affect the ability of the enterprise to win.

What success models do lower-level managers and technicians rely on to guide them toward the right choices? Typically, certain types of choices have been more visible than others. Having seen or heard of such choices being made, everybody knows what is right. Choices that

FIGURE 17–2
Return on Assets

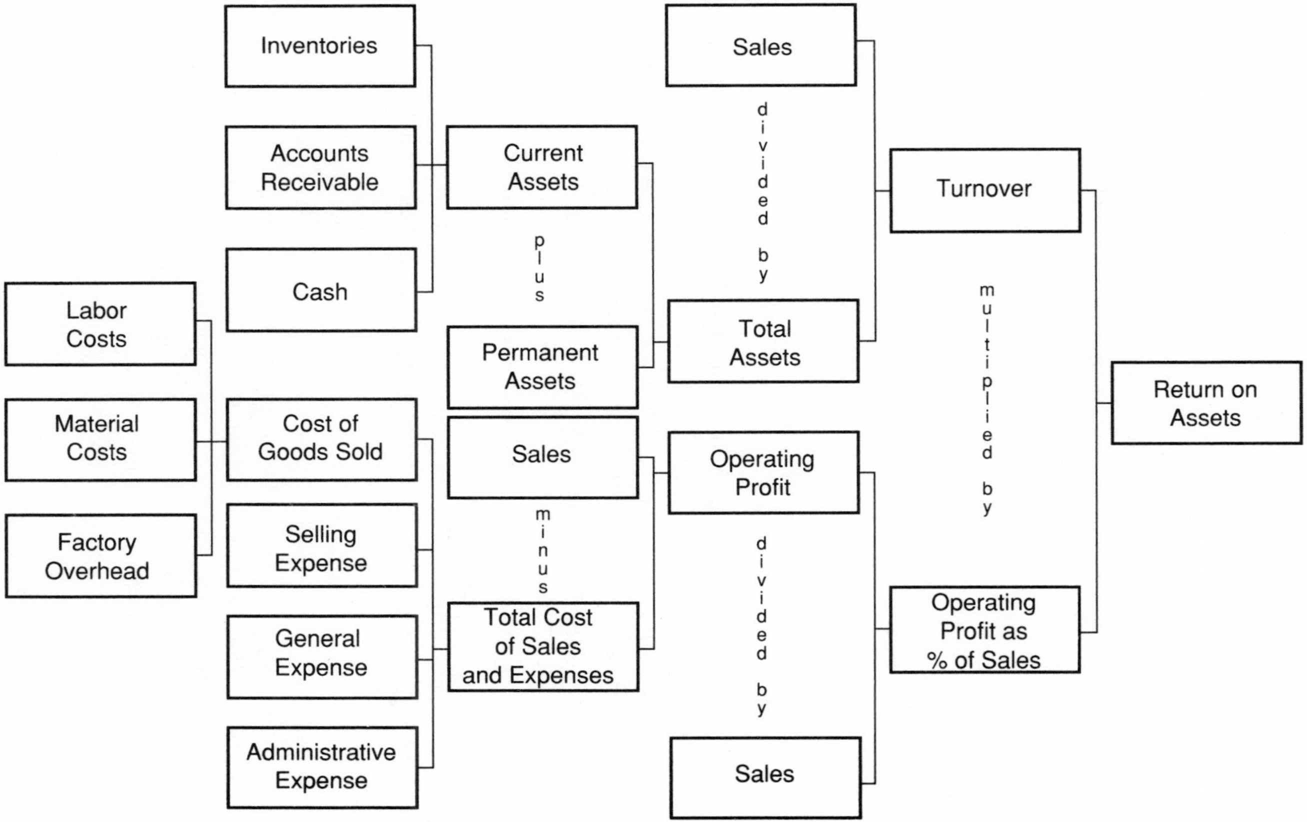

haven't been very visible are the real problems. For those, managers and technicians look to departmental, functional, industry-wide, or even cultural success models for guidance. Therein lies the problem.

Some of those success models are inconsistent; others are just plain wrong. The net result of individuals within the same enterprise relying on inconsistent success models is negative synergy. That saps the enterprise's collective will and ability to compete. Top management actions in such enterprises typically reflect an intense desire to make one or more of the numbers in Figure 17–2 look better. Unfortunately, those same actions often reflect a total lack of understanding of how to accomplish the desired result.

HISTORICAL EVOLUTION

A lot can be learned from reviewing the success models adopted by product-based enterprises in the United States during the past several decades. It is widely agreed that the marketeers were in the driver's seat in the late fifties. Next came the financial-control types. It is not the people who were important, however. Instead, it is the success models they brought with them.

At the end of World War II, the United States had a surplus of manufacturing capacity. Because of widespread defense industry employment and rationing, there was pent-up consumer demand. Naturally, various enterprises were only too happy to convert their manufacturing capabilities to meet that demand.

Eventually that demand was satiated. Selling became harder and harder. As sales dropped, so did ROA and ROE. After all, sales revenue shows up in three different places in ROA (see Figure 17–2). At the same time, a new advertising medium—television—offered one more approach to thousands of American households. No wonder marketing people found entrance to the executive suite so easy.

Promotional campaigns designed to prop up declining sales also increased expenses. Naturally, prices stayed high enough to cover expenses and provide an adequate ROE. Ultimately, it was found that while promotion is undeniably effective, there are limits to how far it can go in offsetting uncompetitive prices. Clearly, something had to give.

Salesmen began to complain about noncompetitive pricing. The financial folks replied that price had to cover costs. Manufacturing management opined that direct labor must not be productive enough. Top management started looking for answers and pulled out the organization chart to find someone responsible for cost. The answer was obvious: It was cost accounting. They had a recommendation, too.

According to cost accounting, what was needed were labor standards and work measurement. It worked, too, for awhile, as it often does when organizations are "fat." Labor costs were brought under control, which eased the upward pressure on prices. Just as everyone got comfortable, sales started singing a familiar refrain: Although they hadn't gone up, prices were too high again. It seems that everybody else had caught up.

After deliberation, top management concluded that the cost problem had been underestimated. Back to the organization chart again! Who understands complicated interactions? Ah ha! The answer was there all along. The production engineers are the real experts in complexity. They knew the answer, too: Mechanize. If direct labor isn't there, you don't have to worry about them being productive. Besides, machines don't take vacations, get sick, or take time off for funerals. That worked, too, for a while.

Eventually, the same cycle repeated itself. Costs were still too high. Once again top management looked to direct labor for the answer. By this time, nobody had any bright ideas. The conclusion was inevitable; high labor costs were hampering the enterprise's ability to compete. The union wouldn't cooperate either. Slowly the answer dawned: Move. Go south where the unions were weak, where the work ethic was strong, and where people would work for low wages.

MORE RECENT PRACTICES

That is all ancient history. What about current success models? There has been so much progress that cost accountants now want computerized labor standards and work measurement. Mechanization is no longer a fashionable idea; now companies are trying automation and robotics. Going south is no longer good enough as businesses move offshore by the dozen.

There has been progress, all right. The tools are fancier, but the success models have changed very little. They were not very effective then, and they are not very effective now, for one very simple reason:

Success models that zero in on direct labor are misdirected. Labor has long since ceased to be a highly significant cost factor in most product-based enterprises. Yet it continues to get an undeserved share of attention from top management, middle management, and technicians.

Productivity is not about how hard direct labor works. It is about how well management manages. Likewise, productivity is not simply a function of capital investment. It is about how well an enterprise uses capital. There are no shortcuts, technological or otherwise. Productivity reflects the cumulative impact of numerous choices (past, present and future) that are made across all disciplines, not just in manufacturing: product, process, procedure, plant, performance measurement, policy, and people choices, to name a few. Above all else, high productivity requires consistency among these choices—a common vision of what it takes to succeed across the enterprise and through time. Without a coherent prescription that recognizes and provides for all the interactions among the seven Ps listed above, managing for productivity and, hence, increased competitiveness is impossible.

Product-based enterprises operating without concise, coherent, and comprehensive success models are intellectually bankrupt, as is their top management. It is time to rethink the whole productivity and competitiveness issue.

TIME, COST, AND INVENTORY

Time and Cost

For too long our perspective on cost has been clouded by the view of individual processors as independent agents. When processors are viewed as user/supplier members of interdependent cycle teams, that perspective changes:

1. Activity takes time and costs money.
2. The longer it takes, the more it costs.

These two points are expressed graphically in Figure 17–3. Note that as long as cost is occurring at a given rate per day of cycle time, longer cycle times mean higher cost *regardless of whether steel or information is being processed.*

FIGURE 17–3
Cost-Time Relationship

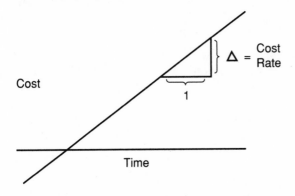

Inventory

There are two kinds of inventory: (1) Highly visible inventory consists of raw materials, work-in-process, and finished goods. (2) Invisible inventories consist of money paid for various process-cycle activities (e.g., payments made to hourly production personnel for material to be purchased next month) in advance of the recovery of those expenses through sales. It can be shown that cycle-time reduction always leads to inventory reduction under certain conditions, regardless of whether the inventory is visible or invisible. In general:

$$\begin{array}{c}\text{Average}\\\text{Units In}\\\text{Process}\end{array} = \begin{array}{c}\text{Throughput}\\\text{Rate}\end{array} \times \begin{array}{c}\text{Cycle}\\\text{Time}\end{array}$$

The number of units processed per time interval must be the same before and after the hypothetical change. The second requirement is that the processing cost rate per day must not increase such that the total cost of processing a single unit is greater than it was before the hypothetical change. If in fact the processing cost rate does not increase that much, or even decreases, reductions in cycle time will also reduce both cost and inventory, since generally:

$$\begin{array}{c}\text{Average}\\\text{Dollars in}\\\text{Process}\end{array} = \begin{array}{c}\text{Average}\\\text{Units in}\\\text{Process}\end{array} \times \begin{array}{c}\text{Average Cost}\\\text{in Units in}\\\text{Process}\end{array}$$

SHORT-CYCLE MANAGEMENT

Clearly, cycle-time reductions increase competitiveness by increasing responsiveness. The previous discussion makes it obvious that not all cycle-time reductions are cost effective. Likewise, some means of reducing cycle time can make quality worse, not better. But the discussion also raises the possibility of simultaneous improvements in responsiveness, visible and invisible inventory, capital requirements, and total cost. Suppose certain principles could be found that not only can be used to do all that for both white and blue collar processing activities but that also improve quality at the same time.

That is exactly the magic of the short-cycle management (SCM) principles. Close examination of the JIT/TQC/TPM (just-in-time, total quality control, total productivity measure) concepts, tools, and techniques incorporated therein suggests that the primary criterion for their selection was their ability to complement one another. In every case where one of those concepts, tools, or techniques has a potentially negative outcome, some other one has obviously been developed and incorporated to offset it. Smaller processing batches and changeover reduction provide an obviously instructive example of this approach. There are many others.

As low-cost, high-quality products become the price of admission to some markets, customers begin to think in terms of total enterprise cost, quality, and responsiveness. These changing market needs are on a conflict course with current and past trends in traditional manufacturing enterprises. Meeting the responsiveness challenge depends on an enterprises' ability to identify and shorten the four primary business cycles: book/bill, design/develop, spec/source, and purchase/produce. Some approaches to shortening cycle times impact cost, quality, and capital requirements negatively. Short-Cycle Management, however, incorporates numerous JIT/TQC/TPM principles selected to complement one another. Those principles can be used to dramatically improve competitiveness by simultaneously improving white and blue collar processing cycle times, cost, quality, and capital requirements.

COMPETING IN GOOD TIMES

Good times are characterized by a generalized sense of well-being. There seems to be enough demand for everybody, and overhead absorption is not a problem for most enterprises. Customer questions focus on when and what quantity, instead of how much does it cost and whether to buy or postpone.

Surviving and prospering in good times is harder than it sounds because demand is not inexhaustible. Enterprises must still compete for shares of limited markets. All an enterprise has to do to win the market share game in the short run is offer the highest perceived value for the money—provided it has the necessary capacity. The enterprise has to offer the highest quality, the lowest price, the most dependability and the quickest response—simultaneously. If any one of those factors slips, competitors are typically poised to take full advantage of the lapse. Chasing all four factors at the same time is tough enough.

As customers become progressively more sophisticated, their expectations are harder to meet, even in good times. More specifically, product quality is no longer the only relevant consideration. They expect quality at every customer-supplier interface. Invoices, shipping documents, instruction books must be right the first time.

Unit price and payment terms are no longer the only issues. Customers want to know how many paperwork hoops they have to jump through to do business with you on an ongoing basis. Many have decided that carrying six months of purchased inventory to get another percentage point or two shaved off unit price doesn't necessarily make sense. The *total cost of doing business with a supplier* is the issue.

When it comes to dependability, customers used to be satisfied if you delivered roughly the right quantity of what was ordered at approximately the right time. Now they expect exactly the right quantity at precisely the right time. If sales, customer service, accounting, engineering, or any of the other folks that come into contact with customers make promises, the customer expects those promises to be met. Customers expect integrity in all promises.

Then there's responsiveness. Performance to promise is no longer good enough. The promise itself has to be attractive. Of course customers want products, answers, documentation, and information delivered as promised. But they also expect those promised deliveries to be a lot quicker than they once did. Sometimes the expectations are for immediate answers; tomorrow is too late. The expectation is clear: responsiveness everywhere.

Making Money in Good Times

There is considerably more to winning the competitiveness game than just capturing market share—even in good times. If an enterprise loses money on every unit it sells, it cannot make up the difference with more share. That is just an extreme example of a more general case. If an enterprise is not making enough profit on each unit it sells, it will not solve the problems with more volume.

Profit margins are clearly extremely important in good times as well as bad. How much is enough? That depends on the size of the asset base, that is, how much money is tied up in equipment, inventory, systems, space, and other capital "sinks." *The more capital tied up, the greater the profit margin requirement.*

Publicly held corporations typically rely on capital markets for asset-based funds. If current profit levels are low relative to those capital requirements, capital may be unavailable, or hard to obtain and expensive. Unavailable capital limits growth, while expensive capital further reduces future profit margins by further raising the cost of doing business.

SURVIVING IN BAD TIMES

Bad times are characterized by widespread anxiety and depression. There isn't enough demand to go around, and most enterprises seem to have trouble absorbing their overhead. Customers are primarily concerned with whether to buy or postpone. Quantity is not a consideration; buying as little as possible as late as possible is a foregone conclusion.

Winning market share in bad times is even tougher than in good times as individual competitors fight for severely limited demand. Who is the winner in these circumstances? Clearly, it will be enterprises with the lowest prices, won't it? Not necessarily. Typically, there is at least one maverick willing to meet the lowest comparable price and still provide the quality, dependability, and responsiveness that customers have grown to expect during good times.

Who are the mavericks? They are enterprises that made unwavering commitments to their customers some time ago. They are enterprises that know their customers, as well as their needs and expectations. Once again, the market share winners will be enterprises that provide superior quality, dependability, and responsiveness at all the customer interfaces at low cost—for a while.

A strong customer orientation is clearly a necessary condition for winning over the long run, but it is not sufficient to guarantee success. The winners during extended bad times are enterprises still on their feet when the bad times are over. The winners will be some subset of the mavericks described above.

What happens to everybody else? During extended bad times, those without a strong customer orientation are typically the first to go. They simply can't keep up in the scramble for the little demand that remains. The rest of such shakeouts appear to depend on how far down the enterprise can follow the inevitable price slide and still make money.

Some Japanese enterprises learned this lesson the hard way during the oil shock in 1972. Previous spending on automation had positioned them for high margins during good times. When the bad times hit, volumes shrank accordingly. At lower volumes, profit margins simply were not sufficient to support that level of capital investment.

When bad times last long enough, an enterprise faces a Hobson's choice. On one hand, they may choose to keep prices high in an attempt to maintain margins. The inevitable result is still less volume. Or, they may choose to follow prices down until they are losing money on every unit. Either choice leads ultimately to the same destination—insolvency.

Clearly, the winners in bad times are the same as the winners in good times: enterprises who demonstrate an unwavering commitment to meeting the expectations of their customers and keeping operating expenses and capital requirements well in hand.

THE CAPACITY CONUNDRUM

Long-term capacity decisions are made in good times and bad times. Enterprises that build excess capacity during bad times are viewed by some as farsighted and by others as foolish. What do those opinions reflect?

Those who see capacity expansions in bad times as foolish recognize that conventional approaches to building excess capacity take capital and raise breakeven volumes. They argue that the future is unknown and that the unknown is threatening.

According to this view, such expansions shorten the time that the enterprise can survive in bad times. Given the belief that extended bad times are inevitable, capacity expansions during such times are clearly shortsighted and foolish.

On the other hand, those who view capacity expansions during bad times as farsighted appear to view the world through rose-colored glasses. What goes down must come up. The enterprise that is not correctly positioned for the good times may miss a once-in-a-lifetime opportunity.

What about capacity expansions during the good times? Experience to date indicates that during such periods, the optimists clearly have the floor. Pessimistic arguments get short shrift during the good times. Witness what has happened to semiconductor capacity over the last couple of decades.

Who is really right, the optimists or pessimists? Is luck the real determining factor? Both are logical arguments. Neither considers the other side of the coin. There is another answer.

Capacity is not purely a question of capital investment. The winners of the competitiveness struggle are typically unusually nimble. They flex volumes up or down to produce the volumes dictated by bad times or good times. Even more important, they do this with minimum capital investment, hence low breakeven volumes.

Shooting stars invest heavily in good times and bad, always trying to stay ahead of the market growth curve. They make lots of money during good times. High margins and large volumes go well together. When bad times come, they falter quickly as volumes plummet. Ultimately, if the bad times last long enough, shooting stars flame out.

Winners make money in bad times and good. During good times, they don't make as much money as shooting stars because their high-volume margins aren't as good. When bad times come, they make more money because their low-volume margins are better.

When shooting stars flame out, winners are often standing by ready to grab their capacity assets at bargain basement prices. Winners expand all right, but the way they go about it doesn't lead to unnecessarily high breakeven volumes.

A SUCCESS MODEL FOR WINNERS

Winners demonstrate a sense of balance not shared by their less effective contemporaries. They seem to have a knack for satisfying customer expectations, making money, and surviving regardless of changing world conditions. Isn't that what competitiveness is all about? How do winners maintain that balance? What do they understand that their less competitive contemporaries don't?

Losers try to manage results; winners manage for results. Losers manage market share, sales revenue, and return on assets somewhat independently. Winners rely on actions designed to increase market share, sales revenue, and return on assets simultaneously. Losers rely on superficial success models. Winners rely on highly robust, multilevel cause-and-effect success models.

The success models that winners use appear to be based on an appreciation of the many interdependencies at work within an enterprise. Such models enable the prediction of the direction and magnitude of the multiple impacts of individual actions by managers and technicians. Consequently, winners are able to screen out bad actions and concentrate on good ones.

Success models that winners use are typically based on performance-improvement drivers. Consequently, potential actions or changes can be evaluated in terms of their likely impact on the market share, sales revenue, and return on asset drivers.

PERFORMANCE IMPROVEMENT DRIVERS

At this point, the drivers of market share and sales revenue seem like old friends. The determinants of an enterprise's ability to win market share were described earlier as quality, price, dependability, and responsiveness. Although it was not pointed out in the discussion of winning market share, it should be self-evident that increased market share is typically accompanied by increased revenue.

What about operating costs and capital requirements, the other components of return on assets? What drives them? Like market share and sales revenue, they are ultimately driven by choices made by managers and technicians, choices made in all disciplines and through time: product design, process selection, procedure development, plant layout, policy adoption, performance measurement definition, and even people choices.

The proximate drivers of return on assets are sales revenue, operating costs, and capital requirements. In turn, the proximate drivers of operating costs are people, materials, services, and time. *People* refers to the total number of employees of the enterprise. *Materials* refers to all raw and consumable purchases made by the enterprise. *Services* refers to everything it takes, supplied internally or contracted externally, to actively support those who serve the customers of the enterprise. *Time* refers to the interval between completion of the activities required to process materials, paperwork, and the like and the initiation of those processing activities.

The proximate drivers of capital requirements are equipment, inventory, system, and space. *Equipment* refers to manufacturing and other capital goods required for continuing operation of the enterprise. *Inventory* refers to visible and invisible inventory: materials and white collar activities that tie up capital. *Systems* reflect primarily the hardware and software investments typically viewed as necessary evils but most certainly the result of choices by managers and technicians. Finally, *space* is required to house enterprise operations. Generally, choices that require more space, services, and so on cost more, raise capital requirements, and thus reduce return on assets.

STAYING ON COURSE

The message is clear. To be competitive, enterprises must watch their performance-improvement drivers closely and avoid clearly incorrect actions. Top management can help by demanding increased rigor in the ways its subordinates think about improving operations.

Current practices require a thorough review of the trade-offs that will result if a proposed improvement is adopted. Unfortunately, trade-off analysis gives the appearance of objectivity to decisions whether they are right or wrong, good or bad. It is clearly a rational approach, but is it really objective? Hard-earned experience indicates that analysts are quite good at identifying and estimating direct costs and benefits, but not so proficient when it comes to the indirect ones.

Some real mistakes could be avoided by asking the right questions before worrying about detailed financial-feasibility analysis. At any one time, an enterprise's competitiveness can be characterized by its market position, the level of its operating expenses, and the amount of capital it employs. The current level of competitiveness represents the cumulative impact of the efforts of numerous individuals working together to get market share up and the cost and capital requirements down.

Having reached its current level of competitiveness, why would an enterprise ever deliberately backtrack on any one of those three issues? If share is surrendered, revenue and future growth are sure to suffer. If operating costs are allowed to increase, margins and return on assets will suffer. Assuming a relatively constant volume, if capital requirements are permitted to rise, return on assets will suffer, there will be increased upward pressure on prices during bad times and the organizations' ability to survive those bad times will be diminished.

Preventing backsliding is not that hard. All it takes is a simple questioning procedure. Is the proposed improvement likely to worsen our market, cost, or capital requirement position? Examination of the potential impact on each of the four performance drivers often makes the answer obvious. If the answer is no, the proposed improvement probably makes sense and warrants detailed financial analysis.

What if the answer is yes? Then the proposed improvement needs to be rethought. Can the proposal be modified so that its good features can be salvaged without any accompanying negative impacts on the other determinants of competitiveness? Once it has been reworked, the modified proposal should be subjected to the same questioning as before. Sound tough? It is. Most proposed improvements fail the first time, many the second, and some never make it. This is the kind of rigor it takes to turn losers and shooting stars into winners in the competitiveness sweepstakes.

Exceptions to the Rules

Like all conceptual models, the one described above is considerably harder to use than it sounds. What if the proposed improvement affects some of the performance-improvement drivers negatively and others positively? The answer is to stick to competitiveness-determinant categories. Within categories, trade-offs make sense as long as the total impact is positive. For example, an increase in equipment or systems requirements might well be justified by more than offsetting decreases in inventory and space requirements.

As long as market share, operating costs, and capital requirements are not allowed to regress, the enterprise should become increasingly competitive. Every adopted improvement affects at least one of those three determinants of competitiveness positively and none of the others negatively. Consequently, the enterprise becomes progressively more competitive. Now that's continuous improvement that counts.

What about technological breakthroughs, volume increases, and the like? Aren't there legitimate reasons to make capital investments? Sure. One exception occurs when an enterprise sees an opportunity for a strategic breakpoint. Investing in certain new technology is seen as affecting one of the four market-share drivers so positively that the enterprise believes it can change the rules of the game in the market. In such cases, the enterprise believes it can capture share, increase volume, and grow revenue by making a capital investment in technology. In that case, the proposal needs to be looked at in terms of what it will do to the total return on assets. If the impact is going to be negative, it is still a highly questionable proposition. If the projected ROA is satisfactory, the impact of the proposal on breakeven volume needs investigation. Will making the investment reduce the enterprise's ability to survive bad times? By how many months or years? Is it worth it?

Still another exception occurs when the competition introduces a strategic breakpoint, effectively changing the rules of the game. Then an enterprise has no choice about whether to acquire the new technology. It must to stay in the game. How much is to be acquired and how it is incorporated into current operations must still be decided. Impacts on all twelve performance-improvement drivers have to be thought through carefully, and the negative ones avoided wherever possible.

SUMMARY

Competitiveness describes an enterprise's ability to survive and prosper year after year, in good times and bad. Managers and technicians make decisions that improve or diminish competitiveness, based on success models that, for the last several decades in the United States, have been decidedly superficial. The inadequacy of these models is the root cause of American manufacturing's competitiveness problems. Top management must encourage performance improvement by adopting a new success model and using it to screen operations-improvement proposals. We can improve competitiveness by sensitizing managers and technicians throughout the enterprise to what determines competitiveness and how to achieve it. By rethinking the competitiveness issue and educating management and the workforce, U.S. producers can accelerate the change process.

PART 6

DYNAMICS OF CHANGE

Part Six, "Dynamics of Change," explores the difficult issues of disruption and culture change in organizations. Dan Ciampa looks at the responsibilities of top-level managers. The Stanadyne case outlines the history of one company's move from a traditional job shop to a just-in-time (JIT) environment, complete with pay-system changes.

CHAPTER 18

MANAGERS AS CHANGE AGENTS

Dan Ciampa

Editor's note

What can managers do to speed effective culture change? In this chapter Dan Ciampa looks at the new role of managers in an environment of empowered employees, where responsibility is pushed down and spread out in the organization. New competitive techniques, such as simultaneous engineering, and the use of power as a motivator are also discussed. Finally, the author summarizes the seven climatic factors that build commitment and the desired competencies that should aid the selection of project managers in continuous improvement efforts.

Change as a central theme is certainly not new. Bennis, Benne, and Chin captured it quite nicely in *The Planning of Change*,[1] as did many others. The manager's role in trying to have some control over the pace of change is not new either, as anyone knows who has picked up one of Peter Drucker's pieces over the years. So, what can be said that is new and different? Well, there are quite a few things.

Let's start with the word *manager*. When Warren Bennis and colleagues penned their essays, the concept of a manager was more limited than it is today. Organizations were more hierarchical than today's are trying to become. While pushing down responsibility for decision making to lower levels was advocated by some visionaries and experiments did take place, they were the exception to the rule, and generally they were not very well accepted.

EMPOWERMENT

Today, on the other hand, there is not only growing acceptance of the idea of employee empowerment, it is being enacted in every corner of the industrial landscape. The question for many leaders is not whether it makes sense to push down responsibility; the question today is how to do it fast so that employees facing problems day in and day out can solve them when they first occur or, better, can anticipate and prevent them. We want to place action at the source, not dependent on relayed direction from someone who may be more remote. This means that workers assume more responsibility and exert a certain degree of influence over their surroundings. It means that workers exercise the power for decision making—that they become their own managers. The challenge for those who are in positions we would ordinarily consider management is how to ensure that this sense of power pervades those in his or her charge.

THE CHANGING WORKPLACE

Similarly, the term *change* in the phrase "managers as change agents" has taken on new meaning. Change has probably always seemed unprecedented to those experiencing it. Look at the mid-1800s. In spite of the Victorian values for which that era is best known, events that truly changed the world were numerous and happened in rapid succession compared to earlier times. Within little more than a decade there were several astounding feats. John Wesley Powell led expeditions down the Colorado River to map the Grand Canyon. Othnial Marsh, not yet 40 and the country's only professor of paleontology, discovered the fossils that helped prove Darwin's theories of evolution. The Roeblings built the Brooklyn Bridge. The Union Pacific railroad was completed. The Suez Canal opened, bringing Europe and India 5,800 miles closer. These were events for which there was no precedent. They caused those who lived in the time to be astonished by their importance and the ever more rapid succession in which they occurred.[2]

While our grandchildren will probably see our rate of change as slow in comparison to their own, we certainly face major changes in the manufacturing world. The most substantial ones have been brought about by the internationalization of the market and, in particular, the competition. The emphasis on quality and on gaining market share through customer satisfaction has ushered in a new way of thinking about the process of manufacturing. We are altering the nature of work itself rather than just developing faster or less expensive ways to do the same things. The extent of these changes is even more significant than their rapidity. We are starting to change the way we do work (the technical elements of work), but *at the same time* we are proactively seeking to alter the work climate in our companies so that people relate with each other differently: more collaboration, more communication, and more commitment to carry through what is decided upon. The most significant factor in the kind of change we face today is that it is multidisciplinary, crossing from quantitative, technical, hard sciences to the more qualitative, behavioral, soft side of the equation and back again.

TOTAL QUALITY

We are seeing developments such as the total quality (TQ) movement, on a large scale, and, simultaneous engineering on a more functional level. TQ is typically a company-wide effort seeking to instill a climate of continuous improvement on a permanent basis toward products and services that customers will find more satisfying. First, this effort includes programs to educate employees to make them more aware of the need for and principles of better quality in everything they do. Second, statistically based techniques are taught so that root causes of quality problems can be more easily identified and displayed. Third, there is often more emphasis on teamwork and cooperation between departments. Training in teamwork usually takes the place of task forces asked to solve specific problems. Fourth, more emphasis is placed on customer needs through surveys and visits of managers, manufacturing workers, and engineers. Fifth, the TQ process reaches back to suppliers as well, as companies forge closer relationships with fewer suppliers—only those who can best meet their needs. While success in TQ efforts requires the leadership of the most senior person, it is equally important that all employees in the organization feel part of the effort and contribute to its success in their own ways.

SIMULTANEOUS ENGINEERING

Simultaneous engineering is a different way of designing and manufacturing products. It brings together design engineers and their counterparts from production at the time the product is being designed so that the product can be made efficiently and so that quality considerations can be taken into account as a design criterion. Problems in production are avoided by manufacturing having input upstream in the product development process. It also paves the way for engineering help downstream while the product is in production. The essence of simultaneous engineering is the relationship that develops between people from different departments and the fact that each is focused on the task of getting the product right the first time. This sort of relationship based on common goals is unusual in the traditional product development process where engineering-manufacturing relationships have tended to be arm's length at best.

Simultaneous engineering takes different forms. In the case of the Buick-Oldsmobile-Cadillac Group Engineering Center in Flint, Michigan, it means that design, product, and process people are one group in a single location. In another case, GE in Lexington, Kentucky, a multidisciplinary team of design engineering, manufacturing, and support people is formed when a new product idea is created. One result of all this is that the product is not designed and made by sequential and separate activities but by overlapping activities, many of which take place simultaneously. So far, this approach has been used within companies and only sparingly between companies and their suppliers; when simultaneous engineering at this level is more widespread, much more substantial gains will be made. The key is common goals, communication, common measures of success, and teamwork.

Both the total quality movement and simultaneous engineering require that technical and people-related elements of change have equal weight and work hand in hand. It may not be the Suez Canal, but for those trying to manage it, it seems as formidable.

ORIGINS AND PAWNS

This brings us to the definition of *agent* as one that acts or exerts power, a person responsible for his acts. In this sense of the word, there is the implication of action, of doing something, of making good use of the changes that occur. In the 1960s, when there was much change and many people trying to figure out how to manage it, there emerged the theory of origins and pawns.[3] It suggested that there are two sorts of managers in the face of rapid-fire change. *Pawns* are those who let change control them, who surrender to conditions over which they believe they have no control. *Origins* are the opposite. They are the managers who have more confidence. They will take responsibility and stand apart from others when change causes most of their peers to revert to pawn-like behavior. In basketball parlance, these are the "go-to" players, the ones to whom you get the ball when the game is in danger of getting out of your control or when it is on the line. They are the ones who fight to get into the right position and demand the pass.

The notion that every employee, not just those with managerial titles, is a manager and the roles of various people (agents) will be explored and described below.

EVERYONE IS A MANAGER

Until fairly recently, many people believed that bosses had the sole responsibility to change how work was accomplished in their department or company. After all, it was he or she who was at the top of the pyramid, who went to corporate at the end of the year to present the unit's profit plan, who had the corner office and the parking spot with his name on it. Employees looked to the boss for answers. The job of the boss was to manage. That meant to take the lead and tell people what to do. One tongue-in-cheek book on managers put it this way: "For one thing, to be a real boss there were few restrictions to contend with: you hired whomever you chose, you told these people what they were to do, you paid them a nominal wage to do it, and you fired them if they didn't do it correctly."[4]

The old assumptions saw the boss as administrator or strongman. Michael Maccoby described them in this way: "The administrator is the traditional expert engineer, accountant, or lawyer . . . he expects organizations to run by the book . . . construct the right structure, provide proper incentives, and the industrial machine will run efficiently." He continues, " . . . the strongman (. . . like a jungle fighter) . . . believes he can overpower distrust and gain accord by bearing down on his subordinates."[5] One of the most significant changes in the business world today has been the very concept of the boss. Old assumptions are giving way to new ones that see the boss as a person with somewhat different skills than were expected in the past. What we have realized is that both the administrator model and the strongman model are not good enough to take control of change in today's environment.

Economic incentives and the comfort of a strict order of things, or structure—the main tools of the administrator—are not the powerful motivational tools they once were. The achievement-oriented worker today wants the freedom to establish goals and the sense of personal influence to attain them; he wants to have input and involvement. Consequently, the strongman misses the mark as well by sacrificing a sense of teamwork for his own dominance. Today's worker sees two satisfiers, equally important to personal influence and input: (1) relationships of trust with co-workers and (2) a sense of a team composed of different but interdependent parts, all moving in the same direction to achieve common goals.

POWER AS THE MOTIVATOR

We are seeing the emergence of a new set of assumptions about managing today in which the person in charge must care about developing the people under him by inviting input and sharing power. Management presents as an incentive the chance to influence operations rather than using the old piecework-oriented incentive method that said, "Do what I want you to do and I'll give you more money." It is not that people today believe money is not important; the point is they are less likely to accept money at the expense of influence over the things around them and over input into the issues that concern them. All this means that the boss must act differently. He or she must allow subordinates to be influential and find a way to ensure they take responsibility for solving problems. The boss must see that positive relationships are built, ones created on the basis of trust and common belief in a vision of the future. The boss must be persuasive in a way that does not diminish trust. The result should be a work environment or climate where there is continuous improvement and sharpening of the skills necessary to compete against ever better opponents.

If these are some of the elements of the boss's new mandate, what is required of the people in his organization? They, too, must become agents of change. They must welcome a new level of responsibility and fulfill its mandate. The simple fact is that people in the middle and lower strata of the organization chart are being called on to solve problems and make decisions. In some organizations today, it is less an issue of bosses not wanting to share power than of creating the sort of climate where people can exercise the decision-making power available.

THE IMPORTANCE OF THE PILOT TEAM

The vehicle by which decisions are being made is the task team or pilot team. For example, a common building block of a TQ effort is project-by-project improvement. Teams of workers are formed to discover the root cause of a problem affecting them and are asked to find and implement solutions. Decision-making responsibility is passed from bosses through such teams, often as part of an overall company-wide improvement effort. This is a phenomenon that is occurring with increasing frequency, and, by all indications, it will continue. For the past several years, larger companies (e.g., Hewlett-Packard, Apple Computer, Xerox, Ford, and IBM) have adopted this style. Some medium-size firms have moved in this direction as well, because their leaders found this operating style more attractive. The team-oriented, push-the-responsibility-down style has picked up more steam as large corporations have demanded that their suppliers adopt methods that require it. Zero defects and just-in-time (JIT) delivery are becoming the norm as large companies build more collaborative relationships with their suppliers. Along the way, those companies are training suppliers to operate in the way they themselves have learned to operate.

What all this means is that as much of a change of style, perhaps more, is being asked of the employees as of the boss. New ways of operating require a number of things to succeed, but the one paramount requirement is the commitment of people throughout the company to make new ways of operating work.

The techniques embodied in new ways of operating more effectively today make two requirements of employees: They must (1) go out of their way to solve problems and do the right things in the right way and

(2) behave in a team-oriented way. To do these, people must want to; they cannot be mandated. People must care about how they operate and how the work gets done, and that is where their commitment comes in.

SEVEN CLIMATE FACTORS OF COMMITMENT

How can a new level of commitment on the part of employees come about? Telling people to act differently won't make it happen and last. Simply offering to pay people more money won't work either. There is no one leverage point, no one answer. A number of things must be done simultaneously. A certain climate or environment must be created in the company in order for new behavior to take hold. Our experience and empirical research at Rath & Strong indicate that there are seven key elements in the climate of an organization that foster employee commitment to new ways of operating and to accepting new levels of responsibility.

   1. *Influence*: The degree to which people believe they have influence over changing what is around them. Resistance to change is higher among people who have a low sense of influence because they have no sense of ownership in the process and little self-confidence about making conditions different. Many mid-level managers and supervisors often believe they have less influence than they should have, given their title or spot on the organization chart. Sometimes people resist change just to exercise the little influence they believe they have.

   People with a low perception of their own influence who are forced to participate in a team-oriented activity or to accept more decision-making responsibility often acquiesce, but they are usually not energetic, persistent, or committed. In order for people to accept more responsibility, they must develop more of a sense of influence and control, as well as confidence that they can be successful at making decisions they once depended on others to make.

   2. *Responsibility*: Willingness to assume responsibility for doing the right things the right way. People who have a healthy sense of influence are comfortable accepting responsibility. Those with a low sense of influence are leery of responsibility, often believing it is a way for bosses to "dump" more work on them. Companies that have a history of compartmentalization, where engineering hardly ever talks with manufacturing and neither department relates much with sales or marketing, often have a low collective sense of responsibility across the whole organization. This sort of firm has difficulty successfully implementing change efforts requiring teamwork across department lines, including cross-functional task teams and pilot teams.

High levels of responsibility should correlate with high levels of innovativeness, a desire to change the status quo, and teamwork among workers who are willing to take responsibility for their ideas. Those who feel a sense of influence are usually also willing to push for changes they see as necessary. If one is comfortable taking responsibility, he or she is often willing to work with others to enhance the responsibility of a group.

3. *Innovativeness*: The degree to which new ideas that can improve how the company operates are encouraged and the degree to which they are listened to and considered seriously. Highly innovative companies get ideas from many different locations in the organizational hierarchy. This tends to happen in organizations where four conditions exist. First, where people are encouraged to experiment, they will try new and different ways of operating, knowing that mistakes for the right reasons will cause positive feedback. Second, innovation is encouraged where people are accustomed to operating in small teams or task forces. Third, those who come up with the best new ideas frequently should be held in high esteem and rewarded. Fourth, the leader should encourage people to be constantly innovative.

A climate of innovation will allow people to be different and to question the status quo. It exists in the company where employees believe that the leader wants to find new ways to solve problems and will support new ideas.

4. *Desire to change*: A healthy level of dissatisfaction with what exists, a desire to change what is and make it better. In companies where the desire to change is low, there is typically difficulty developing the sense of influence and responsibility necessary for employees to take on more of the decision-making load. At the same time, if the desire to change persists without being responded to for a significant period of time, it can turn from healthy dissatisfaction to unhealthy frustration. When employees have crossed the line from dissatisfaction to frustration, the likelihood of constructive change through employees working together and taking on more decision-making responsibility diminishes.

The desire to change often occurs when there is a crisis. Digital Equipment Company, the computer giant, is a case in point. The first quarter of 1984 was the first time DEC lost money; the stock price fell 30 points. This was the slap in the face that roused its employees and caused them to search for better ways to operate, particularly in manufacturing. DEC's continuous improvement (TQ and JIT) effort resulted in record growth in volume and profits by 1989. While DEC's senior managers contend it would not have happened unless the company had been in a crisis, the challenge they face today is how to maintain that spirit and sense of urgency, that desire to change, during a time of success. The American attitude, "If it ain't broke, don't fix it!" is perhaps the biggest barrier to ensuring success without a crisis. Companies that foster a high desire to change tend to avoid such crises.

5. *Satisfaction*: The degree to which employees are basically satisfied. This means that in order for people to accept responsibility, exert influence, and become innovative, they must be sure that their basic needs are met—physical, economic, and psychological. The psychological needs that must be met as a company moves to become more competitive are the individual's need to achieve and to be recognized for achievement, the need to identify with some social group (such as a work team or ad hoc task team), and the need to influence and have some control over what is affecting his or her satisfaction. When these basic needs for achievement, affiliation, and power are met, the base is established for individual motivation. Meeting these needs helps provide the motivation needed to make employees go out of their way to improve what is around them, even if they are not specifically asked to do so.

6. *Teamwork*: The working state where people trust each other enough to work together to get a job done. In an atmosphere where there is teamwork, several elements exist:

- Issues are confronted directly, often in group meetings, and the group will stick with the issue until it reaches resolution.
- The group participates in decision making, and in so doing group members are sensitized to each other's needs.
- Feelings are freely expressed, and they are listened to and appreciated by others; motives are open and clear.
- Responsibility is distributed, and people believe they are all in it together; because of this, help is freely offered.
- Power and credit are shared. There is sufficient respect for each member's contribution that it becomes unimportant who receives credit.

• Achieving consensus is easier because people are more of one mind about what must be done. There is mutual respect for others' contributions.

7. *Common vision*: There must be a vision created by the top person and his closest advisors painting an image of the future that is compelling and exciting and around which employees at all levels can rally. This vision not only must be clear and compelling, but consistent and consistently presented in terms that are most relevant to those at each level of the organizational ladder. It also must be based on values that are held as important by employees and to which they aspire.

It must start with agreement by top management on the values that will form the basis for the way the company will operate under TQ and JIT. TQ and JIT will not succeed if the people at the bottom or middle of the organization are operating on the basis of an idea that is dramatically different from that of the leader.

THE CHANGE-AGENT ROLE

These are the elements of the organizational climate in which a team-oriented effort can foster the sort of responsibility employees need to assume. In addition to each and every employee assuming more responsibility for change and becoming a change agent, there are specific jobs to be done in today's corporation that call for new skills. One example is the role of the leader of a company-wide continuous improvement effort. More and more U.S. corporations are attempting to improve by way of efforts variously called total quality commitment (TQC), continuous improvement, resource management, and TQ/JIT. In many, an individual is named to be the project leader, often a former line manager. While certainly similar to those taken in the past by people asked to implement new information systems, pay systems, and the like, this role also has some unique aspects. It is explored here because it is an example of how many companies are sharing decision-making power and spreading the responsibility for managing change.

Choosing a Project Leader

For many companies, the most obvious temptation in choosing a project leader for a company-wide improvement effort is to pick someone for whom co-workers feel a sense of loyalty but who has little else to qualify him for this particular task. This is the wrong person, because the project leader should be a partner of the leader and the top team and because together with them he is responsible for pushing changes down through the organization. Although it is necessary for leaders and managers to carry the message to employees, the project leader must coordinate the effort and make certain everyone has the right information at the right time, and he must troubleshoot and solve problems as the process unfolds.

The project leader must be flexible and able to work in a constantly changing environment. It is imperative that the project leader be someone who is not tied to the old ways of doing business with a vested interest in seeing that one particular group within the organization gains influence disproportionately. This person must also be willing to stay in the project management position for as long as it takes to get the job done (probably 12 to 24 months), but not any longer. He should not use this as a stepping stone to a higher management position and should not try to make the position a permanent one requiring a staff, budget, and ever-increasing funding. His loyalty should be to the company head, not to other managers in the company, although the project leader must be able to get along with members of top management, and they in turn should view him or her as a valuable resource.

In one case, a client took a somewhat unusual step that worked quite well. The company had just completed a reorganization and a major cost-cutting effort. The leader did not want to pull a current manager out of a job, fearing that this would weaken an already severely trimmed-down staff. The step he took was to persuade a man who had recently retired after a successful career to return as a consultant and take responsibility for this effort. This person had no axe to grind and could not be accused of political motives. He had been well-liked and respected. He constantly pushed away opportunities to collect a department under him. What he really wanted to do was finish the job and return to the farm he had bought.

Finding this person reinforced the fact that the project leader's job was a temporary one. The continuous improvement effort went smoothly. After about one year he went back into retirement, and line managers assumed responsibility for the effort.

Having a project leader can make the difference between success and mediocrity in a company-wide improvement effort, but not everyone can do this job well. It takes a person who is able to balance a number of demands. The following list (Figure 18–1) shows the characteristics that seem to make the difference between average and excellent project managers. When it is decided to have a full-time person as project leader, candidates should be measured against these 22 characteristics.

SUMMARY

We have explored the change-agent responsibility of the boss, the subordinate, and the project manager. Its central tenet is that no one person can manage strategic change successfully, that it takes the collective effort of people at every level of the organization, each doing his own part. It also suggests that there are certain facets of the organizational climate, the work environment, that encourage the sort of behavior needed to bring about positive strategic change.

Ultimately, it is the leader who must ensure the success of such an effort. In doing so, there are several rules (actually the five "don'ts" of leading a company-wide change effort) for successful, strategic organizational change that must be kept in mind.

1. Don't believe that yesterday's answers will work on today's questions. History can provide valuable lessons only after the leader has formulated a new way to proceed. Being tied to the past will often lead to replicating actions taken in a different environment for different reasons by different people—actions that may not suit today's realities.

2. Don't believe that people who have done things one way for years will change their behavior easily if at all. The wise leader will become dissatisfied with the status quo *first*, before subordinates do. He will think about it, worry about it, ask questions about it, and will formulate a way to address it. Others will be slow to grasp the significance. While the leader may not necessarily drag others along, they will feel the sense of urgency only after he does. Some may never feel his sense of urgency. When it is time to act (crunch time) the leader must be realistic in the way he pulls the organization along, giving people enough time to get to the point of urgency. At the same time, he must not hesitate to replace people, especially high-level people, who will never achieve the sense of urgency enough to change their behavior. Termination is not an easy decision and should be a last resort, but it is sometimes necessary.

FIGURE 18–1
Competencies of Project Manager for a Continuous Improvement Effort

1. Able to implement without taking credit.
2. Able to organize work and bring about order from chaos.
3. Able to set priorities effectively.
4. Able to keep priorities clearly in mind.
5. Knows when to push and when to wait.
6. Gains satisfaction more from goal attainment than from teamwork.
7. Gains satisfaction from knowing things he is responsible for are under control.
8. Thrives on predictability and structures things he is responsible for so they are predictable.
9. Able to traverse organizational boundaries smoothly, and is equally effective in different organizational settings.
10. Has influence skills developed to the point where he can be equally effective with different senior managers whose styles vary widely.
11. Able to listen actively so that the other person feels comfortable sharing information.
12. Able to figure out how to structure an effort so that people from many different camps feel involved.
13. Has loyalty to the leader and to his vision of how the TQ/JIT effort should be structured.
14. Is much more attuned to short-term goals than to a more nebulous, long-term vision.
15. Cares more about getting things started than about carrying them forward for a long time, and is happy to pass off responsibilities as long as the goals he has set out to achieve are attained.
16. Works in a deliberate way and makes steady progress, even if some see him as plodding.
17. Works iteratively, in a building-block fashion.
18. Automatically breaks a long-term task into shorter-term steps that are more easily envisioned.
19. Does not care about building a staff, having a title, or creating a power group.
20. Buys into the TQ/JIT philosophy and believes deeply in its precepts, but is not dogmatic.
21. Is more solid and dependable than flashy and exciting.
22. Is more practical than ideologically pure. Searches for ways to negotiate and move around obstacles. Is satisfied with having gotten 80 percent of what is needed instead of stubbornly holding out on ideological grounds and getting nothing.

3. Don't believe you can do it all yourself. There is a fine line between confidence and megalomania. The changes needed in most manufacturing companies are too enormous, and the problems to be solved too deeply rooted, to be resolved by any one person. Everyone has a role to play. The wise leader will find a way to have those roles defined and make sure people fulfill the requirements of each. The most effective leaders are those who understand what they do well and do it but who find others to do the things they don't do as well. The industrial graveyard is filled with epitaphs of companies that had a better mousetrap but also a leader who could not or would not bring in people to participate in the change, people who would make the firm a bigger, better competitor.

4. Don't believe you can anticipate everything, or that you don't need any game plan at all. This is related somewhat to Rule 3. There are some leaders who try to anticipate everything that could possibly happen. It cannot be done, and it is the wrong way to think about it; we're talking about changing people's *attitudes* and *then* their behavior. Attitude change is something that cannot be entirely systematic; the comforting sequence of the scientific method can get in the way rather than ensure success. Changing attitudes is iterative at best; not only must one take a step back to repeat the one just taken, but one must often take a step to the side also.

At the same time, a game plan is essential. The right level of planning here is more akin to a basketball team than a football team. In a basketball team, there are only a handful of basic plays. The point guard calls a play as he brings the ball downcourt, but what actually happens depends on how the defense looks, who is overplaying, who gets double-teamed, and so forth. A football team has a play book three inches thick; the players take time to memorize the spot to go to. When a quarterback overthrows a receiver by ten yards, it is usually because one or the other had the wrong play in mind; the quarterback throws to a spot and the receiver runs a particular pattern to a spot. Organizational change is a basketball game, not a football game.

5. Don't believe the old adage, "If it ain't broken, don't fix it!" This flies in the face of every principle of preventive maintenance, from yearly physicals, to getting one's teeth cleaned, to a 5,000 mile checkup on your car. The wise leader is always looking for opportunities to make things better before they break. The cost of repairing a broken organization is enormous compared to the fine tuning and constructive change required to keep it sharp.

NOTES

1. Benne, Kenneth D., Bennis, Warren G., and Chin, Robert, *The Planning of Change* (New York: Holt, Rinehart & Winston, 1968).
2. This sort of passive social change has been pointed out in a much better way by David McCullough in *The Path Between the Seas* (New York: Simon & Schuster, 1977).
3. I have always attributed this theory to Dick Descharmes who wrote a paper on it in 1968 at St. Louis University.
4. E. Nevins, *Real Bosses Don't Say Thank You* (Nevins Publishing Co.: Somerville, NJ, 1983), p. 12.
5. Michael Maccoby, *The Leader* (New York: Ballantine Books, 1983), Preface, p. xiv.

CHAPTER 19

CASE STUDY: STANADYNE, CHANGING THE MANUFACTURING STRATEGY FROM JOB SHOP TO JIT

William G. Holbrook

Editor's note

The Stanadyne case illustrates tactics required to meet the manufacturing strategy of competing on cost, quality, and customer service in the world market. Stanadyne implemented major culture changes in its two-phased restructuring from a classic job shop to JIT and work cells. All pay plans and performance measurement systems were redesigned to match the system changes.

Added to the dynamics of implementing major culture changes, the company restructured and went private shortly after the changes described in the case. The resulting smaller, but leaner and more focused, business is more profitable.

Stanadyne Diesel Systems is part of a 100-year-old company that started supplying screw machine products in the Northeast to various companies, including Pratt & Whitney and Xerox. The company was a job-shop, considered to be a "quality house" for screw machine products.

Forty years ago, the company began manufacturing a proprietary product based on a unique design for diesel fuel injection. The fuel injection business grew and prospered while coexisting with the job-shop business. In 1976, Stanadyne was contracted to supply diesel fuel injection equipment to a large automotive company. Required volumes were so significant that the job shop business was abandoned and manufacturing facilities were converted to production of diesel products only. During a six-year period, sales doubled three times. Facilities grew from one plant to four, one in Connecticut and three in North Carolina.

PRODUCTS

Stanadyne Diesel Systems Division supplies three major product groups to most U.S. diesel engine manufacturers. The customer base is small but well-known. The volumes and mix of business are generally predictable. Lead times from initial product design to shipment are one to two years. Sales are not off the shelf; orders are manufactured to meet customers' engine-build schedules.

The three product groups are:

1. Pumps: The basic component in the fuel system is the injection pump. Containing about 150 components, the major parts are manufactured in Windsor, Connecticut, then shipped along with purchased components to Jacksonville, North Carolina, where they are assembled, tested, and sent to the customer. Volumes run in excess of 1,000 assemblies per day.
2. Nozzles: The nozzle injects fuel from the pump into the engine combustion chamber. It is made up of 15 components manufactured at Windsor and shipped to Jacksonville for assembly and test. Nozzles are currently being manufactured at a rate of 7,000 per day.
3. Filter Products: The Washington, North Carolina, plant manufactures a variety of filters and fuel separators. A filter or separator consists mainly of purchased components assembled and tested at rates approximating 15,000 per day.

FACILITIES

The Windsor plant, the center of all product development and engineering, manufactures most of the components for the pump and nozzle. This site employs about 1,100 people; one-half are involved in production, the rest in engineering and administrative functions. The North Carolina plants employ about 300 to 400 people at each site, devoted almost entirely to production.

THE PROBLEM

In 1981 Stanadyne faced a challenge: how to profitably survive in business with large capacity and overheads in place but with sales shrinking. Business had fallen off very suddenly after a period of rapid growth. The economies of diesel-powered automobiles became less attractive; in 1981 volumes were cut in half. The company was faced with some major decisions about how to exist profitably in the face of decreasing domestic market demand.

THE RESPONSE: A NEW STRATEGY

The clear direction taken in response to decreased orders was to market fuel injection products overseas in Europe and Japan. This strategy presented more questions, however. How could the company do it and still maintain profitability? Prevailing overseas prices were at or very near direct cost to manufacture. Stanadyne was faced with either making some very substantial reductions in manufacturing costs or changing the type of business it was in.

In order to succeed, the company had to define tactics that met these strategic objectives:

1. Improve quality to compete in the world market.
2. Deliver products when customers want them and in the quantities they need.
3. Continue the R&D effort to bring new and competitive products to market.
4. Drastically reduce overhead structures.
5. Reduce the cost of manufacturing.

Phase I: JIT

Fortunately for Stanadyne, the problem appeared at the same time some very forward-thinking people in this country began to investigate JIT (just-in-time) manufacturing techniques to achieve:

- High quality levels.
- On-time delivery.
- Ever-improving cost to manufacture.
- Low overhead, low inventory.
- People involvement.

A philosophy was formed for the business incorporating these five objectives while identifying the most effective approaches for the environment existing in Stanadyne at that time.

The new manufacturing strategy was based on three requirements:

1. *Culture change.* The new operating philosophy had to be part of our culture. There was no time to bring about changes through a typically lengthy project implementation with a project leader and pilot run. The change had to involve every facet of the manufacturing operation.
2. *Continual improvement.* The overriding theme had to be one of commitment to continual improvement and change. It was not adequate simply to meet a specific quality level or to meet the standard. There had to be a willingness to try changes without fear of making a mistake. People needed to agree that if a change did not work, they could come up with a better idea and try again.
3. *Simplicity.* We had to simplify the complexities of processes and equipment that had been put into place in the late sixties and seventies. During that period, we had purchased specialized equipment to accomplish many complex processes. We had also put complex systems into place for financial reporting and material control (i.e., material requirements planning and standard cost-variance analysis). We needed to reevaluate our complex processes and systems in order to reduce the cost of doing business.

Using these three general parameters—the need for permanent culture change, continual improvement, and simplicity—we developed new tactics. We wanted to take the manufacturing facility out of the job-shop mentality and into a mindset dedicated to producing quality products on time and at a lower cost through repetitive processes (see Figure 19–1).

FIGURE 19–1
Stanadyne's Just-In-Time Manufacturing Strategy

Process Design
  Common blanks
  Common process—70%
  Process capability
  Process control—SPC

Manufacturing Flow
  Eliminate job shop concept of shop orders
  Phase out storeroom checkin/checkout
  Smooth the flow

Employee Involvement Plan
  Quality circle program
  Revised incentive pay plan

Material Handling
  Physically link machines—Tracking
  Standard quantity containers
  Reduce distance traveled

Machine Cells
  Dismantle general machining departments
  Develop cells by *family* of parts

Scheduling
  Daily rates of production by family
  Kanbans when logistics dictate
  Daily performance reporting
  Mix schedule through "MRP"
  Reduce role of production control

Lead Times/Inventories
  L/T as a planning tool to control inventory reduction
  L/T does not exist for lines and kanbans
  Inventory levels fixed by line design

Process Design

The processes were redesigned to give commonality through at least 70 percent of the operations for each major family of parts. Each component started out as a common blank and traveled through a common process. Operations to make each part unique were performed in the finishing operations.

After completion of the process design, the capability of each process was reviewed to ensure that the specifications called out on the blueprint could be met.

Process Control

When the process was in place and determined to be effective, the next step was a control mechanism to ensure that the process would remain in control. Statistical process control (SPC) was implemented for the key characteristics of each component group. As SPC was implemented, large numbers of inspectors were eliminated. Responsibility for quality was refocused on the foreperson, who became solely responsible for parts quality. Because SPC proved that the process was yielding good product, final and in-process inspection were eliminated.

We learned very early in the program that for the plan to be successful, the total work force had to be involved. The plan would not have worked if we had tried to merely engineer the change. Because the factory employees were the ones who knew a great deal about process problems, they were brought into the problem-solving mechanism to make the improvements workable and effective.

We started massive training programs to implement a meaningful SPC program. "Quality circles" were instituted in all the plants to teach team problem solving.

Pay Plans

Incentive plans to change the method of paying employees were needed to foster teamwork. The individual piecework plans that had been appropriate for the job-shop environment were changed to group incentive plans. Later, the group plans were changed to straight daywork. The changes in the pay plans were the most traumatic part of the total transformation. The impact was felt not only by machine operators, but by supervisors who had to alter the way they ran the various manufacturing centers.

Material Flow: From Job Shop to Cells

In a classic job-shop environment, similar processes are grouped together. Parts move from one process to another based on a traveler or route sheet. Parts are batched, issued against a shop order, and expedited through the shop to meet a due date. This type of process requires large expenditures for material handling and expediting (see Figure 19–2).

In order to eliminate the lengthy and costly process of moving parts in and out of a queue for each operation, production equipment was reorganized into cells. Each cell performs all the operations on a specific family of similar parts. The result is significant reduction in the distance a part travels through various processes. Machines are physically linked

FIGURE 19–2
Yesterday's Routing

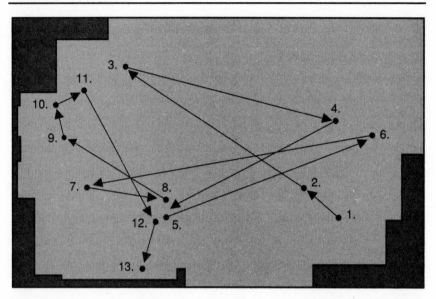

together by conveyors or tracks, using standard containers and constant
container quantities (see Figure 19–3).

Shrinking Production Control's Role
Once the cells were set up and the various processes were balanced to
allow a flow of material, the production control process was reduced
to merely setting the cell's daily-run rate and issuing a product-mix
schedule. The need for expediting disappeared because the location of
every part became known. Scheduling is done for the entire cell, not for
each operation within the process.

Kanban
A system of Kanbans was implemented to shuttle parts between operations. A predetermined number of standard containers (unique for that
component) were used as Kanban to maintain flow without a complex
scheduling mechanism. In the Kanban system, the supplying operations
run until all the containers are full; then the operation is shut off. The
using operation returns empty containers to the supplying operation as
the parts are consumed in the process.

FIGURE 19–3
Today's Process Routing

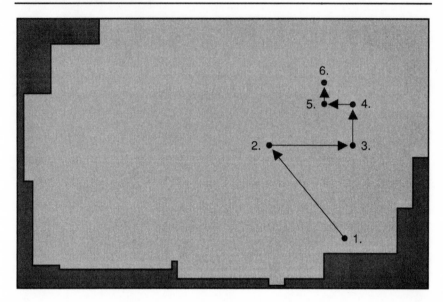

Results

The results were dramatic. With the creation of cells, the entire material control system became very simple. The information needed for control is merely a comparison of planned daily-run rate versus actual production. Cost control, which will be discussed in greater detail later, also became very simple. Lead times were typically reduced by 75 percent, many from weeks to hours. Inventory levels were reduced. There was no need for queues or stock rooms because material continued to move through the process until it was completed.

The reorganization greatly reduced manpower in the factory. Production control slots were reduced from 32 to 4. The material expediters were no longer needed. In their place, four production planners prepare the shop schedule for components produced within the daily-run rate plan.

Quality control has been reduced from a force of over 100 inspectors to a group of 9 people who do in-process quality checks. This reduction was accomplished by concentrating responsibility in the production cell.

Many other supporting departments were downsized by manufacturing process simplification. Areas that were *not* reduced were manufacturing engineering and industrial engineering. All the process and pay-plan changes required a great deal of engineering help. Moreover, it became very clear that the only way to make so many concurrent improvements in the manufacturing process was with the help of production workers, using engineers to support them. Stanadyne's success has been the result of production workers buying into the new philosophies and management letting them work on the problems.

Other achievements have been in the area of work-in-process (WIP) inventory reduction. As the cells were formed, WIP was reduced by 50 percent. A more significant achievement is that the factory now produces on schedule. Although this is a recent accomplishment, it is important because quality improvements or cost reductions cannot be effective until operations are on schedule. When the factory is behind schedule, all efforts and focus are on getting the parts through the shop. There is little energy left to pursue better quality and correspondingly lower cost.

Phase II: Performance Measurement

After we had made many practical changes to our operations, we realized that the reporting and measurement systems were not responsive to manufacturing's new tactics. We needed a new measurement system. Further, before any measurement system could be established, we had to satisfy these prerequisites:

1. *Quality at the source.* We had to make the product right the first time before any changes would be effective in the manufacturing measurement system. Without commitment to quality, it would be pointless to measure an out-of-control process. A system to measure costs and performance of product as it moves back and forth for pickout, rework, and reinspection would be far too complex.

2. *Manufacturing cells.* A manufacturing process organized into cells dedicated to the production of a similar family of components has a very real opportunity to simplify the total-cost and performance measurement system. By being able to focus the measurements for a family of parts onto a single set of resources (men and machines), a simple set of measurements can be established.

3. *Business plan and critical success factors.* In order to measure progress effectively, we needed an effective strategic business-planning process. There was no need to measure something if it had no real impact on the success or failure of the business. The best example of this is the traditional enthusiasm that business has had for measuring the efficiency of labor. We decided to measure factors that were key to the success of the business.

The critical success factors are factors that are so important that if any one of them were out of control, the overall success of the business would be in jeopardy. Not more than five or six factors should fall into this category. The real test of their criticality was to ask, "If I meet my performance objectives on each of the critical factors, will the business be a success?" If not, then that factor did not justify expending resources.

The critical success factors change as the business changes. We review them periodically.

Measurement system design.   Creating the measurement system can be overwhelming, but if addressed properly, it can be simple and easy. After we identified the critical success factors, we attacked the design of appropriate measurements. We felt performance measures for manufacturing companies competing in the world market should follow these priorities:

- Quality issues get top consideration; if the quality of the product is out of control, all other measures become meaningless.
- Second, measure delivery performance to the *customer's* demand. This requires meeting the ship schedule.
- Having met the first and second criteria, the organization should begin to concentrate on the third priority, cost.

The Universal Measure.   Although measurements vary over time and from one business to another, one that is fairly consistent for all business is the actual unit cost.

A universal measure for manufacturing operations makes sense because it is simple and easy to understand. The universal measure is simply "the actual unit cost of production." In a cell dedicated to the production of a family of components with a uniform flow of material (all good JIT practices), it is easy to organize data collection to calculate actual cost to produce a single component. The actual cost per unit can be understood by almost everyone who ever purchased a loaf of bread, a gallon of gasoline, or a bottle of beer.

The costs that go into the equation are the sum of direct labor and fringe, expendable tooling and supplies, maintenance supplies, and direct charges for maintenance and repair orders, divided by the pieces produced. Manufacturing supervision knows that they cannot make more than what is on the shop schedule, so they must manage spending relative to that volume.

Concentrating on actual unit costing focuses attention on actions that the foreperson can control (i.e., his own spending). Historically, business has concentrated on variances from standard. Discussions always bog down over the adequacy of the standard rather than on the actual spending pattern.

The new system now meets all these criteria:

1. It is easy for all employees to understand.
2. It fits all levels of production and products.
3. The results are controllable by the people on the shop floor.
4. It supports the theory of continuous improvement.

Unfortunately, many organizations have been far too concerned with measuring costs and have ignored the first two criteria. This generally happens when an organization becomes a slave to traditional financial measures and does not allow creative design of a measurement system that is truly responsive to the real success factors in manufacturing.

Design Recommendations

There are a few basics that I would offer to anyone who is contemplating redesign of a measurement system.

1. *Maintain flexibility.* Start out with a personal computer to experiment with your measurement systems. Success factors change as businesses grow and mature and as business conditions change. Programming your measurement system first on a mainframe or mini would result in extensive reprogramming and frustration. Also, after factors are identified the first time, it may take several attempts to successfully choose the true factors that guarantee success.

2. *Keep the measurements simple.* A complex system of measurements only understood by the designer is really no measurement system at all. JIT is based on simplicity of design and execution; its measurement system must also be simple and easy to understand. A system that is too complex creates confusion about what must be done to improve overall performance.

3. *Pinpoint responsibility.* Merely stating that a particular factor falls short of expectations is not enough. The measurement system must be capable of tracing back to the operation or function responsible for the shortfall. For example, an indicator measuring quality failures at final-product test is fine as long as sufficient detail exists to be able to trace back to the source of the product failures. Without this capability, the system merely predicts failure without hope of preventing it.

LABOR ISSUES

Compensation Methods

There are a number of issues to be addressed when deciding how much effort should be placed on the control of labor. We decided that, if labor were a small portion of the total cost of production and the policy is to not reduce manpower even if it is not 100 percent utilized, then a complex system of labor control would make little sense. If labor costs are significant, however, then some system of control is necessary. The danger here is spending more money to control labor than will ever be realized from its control. We asked ourselves the following questions:

- Is labor a small portion of direct cost?
- Is labor a scarce commodity?
- Can labor be controlled in a cost effective manner?
- Is there an application for incentive pay?
- How does gainsharing fit into the pay system?
- Are labor standards necessary?

Advantages of Dayrate Plan

In a JIT environment with manufacturing concentrated in cells, individual piecework incentive pay plans run contrary to the objectives of quality and schedule. An operator will not concentrate on quality if his pay is based on additional production. Nor will he produce only what is needed when he makes more money for each additional piece produced.

A dayrate pay plan eliminates the contradictions of an incentive system but creates some other problems. Some workers feel that the opportunity to excel has been removed when piecework is eliminated. The production supervisor's role changes dramatically because, as far as a worker's productivity is concerned, an incentive system is a self-policing mechanism. But a dayrate system requires close supervision of the productivity within the cell. The supervisor must investigate each time production falls below acceptable limits, and he must determine why. The worker who is not on incentive may be less motivated to tell the supervisor that there are problems.

The real solution to this problem is not to make the supervisor into a superhuman individual. It is better to motivate each operator to be more involved in continuous improvements within the work cell. This can be done through some form of gainsharing with the employees and by making the employees aware of the company's objectives and the progress being made toward meeting those objectives.

To summarize our experience in changing the pay plans, we had a few problems, but overall the results were appropriate to the new manufacturing strategy. In 1982, we changed from individual to group incentives for all manufacturing cells. We also changed to dayrates, and found that it works well in cells. It is adaptable to bar coding, and it allows for emphasis on quality rather than quantity.

The problems we encountered converting batch departments to dayrate included losing a worker's incentive to excel or to fix problems. The work pace tends to lead off at 100 percent. Also this method requires accurate standards and voluminous data collection. The answer to these problems is to measure differently, looking at operator productivity and overall department labor utilization.

I have briefly reviewed the points that were important at Stanadyne in setting up a responsive measurement system. It was really very simple; we knew our manufacturing goals, and we designed various check-points along the way to make sure we got there on time and in good condition. I think the discussion here applies to most manufacturing concerns in the United States today that count on profits as their ultimate long-term measure of success.

About the Authors

Sara L. Beckman is the manager of Hewlett-Packard's (HP) Manufacturing Strategic Planning function, where she coordinates the development of the corporation's manufacturing organization strategy. In this capacity, she is involved in decisions regarding degree of vertical integration, determination of plant focus and size, and facilities location. She is also a member of the Business School faculty of the University of California, Berkeley. She represents HP on the Advisory Board of the MIT Leaders in Manufacturing Program.

Prior to joining Corporate Manufacturing at HP, she worked on the development of the Data Systems Division manufacturing and computer-integrated manufacturing (CIM) strategies. She joined HP after four years at Booz, Allen, and Hamilton in their Operations Management Services consulting practice.

Dr. Beckman earned her bachelor's, master's and doctoral degrees in Industrial Engineering from Stanford University where she also received a master's degree in Statistics.

Joseph D. Blackburn is Associate Dean for Academic Affairs and Professor of Operations Management in the Owen Graduate School of Management, Vanderbilt University. His consulting and research work covers operations strategy, material requirements planning (MRP) systems, and new product forecasting models. He has published numerous articles in *Management Science, Marketing Science, Journal of Operations Management*, and other journals.

He is the Director of the Operations Roundtable and an associate of the market research firm Yankelovich, Clancy-Shulman. He also serves on the Editorial Boards of *The Journal of Operations Management* and *Corporate Strategies*.

He received his B.S. degree from Vanderbilt University, an M.S. degree from the University of Wisconsin, and a Ph.D. in Operations Research from Stanford University. He has also taught at Boston University, the University of Chicago, and Stanford University.

William A. Boller is the General Manager of the Industrial Applications Center at Hewlett-Packard. Prior to holding this position he formed the Strategic Manufacturing Consulting Group. Boller was the manufacturing manager for the Data Systems Division at HP. He concurrently served on the Corporate Manufacturing Council as the Group Manufacturing Manager.

Mr. Boller has a bachelor's degree in Mechanical Engineering from Stanford University, a master's degree in Mechanical Engineering from the University of Southern California, and an M.B.A. from Stanford University.

Richard S. Bombardieri is a partner in the consulting firm, Manufacturing Excellence Action Coalition (MEAC). He has 20 years of line and consulting experience in materials and manufacturing management; he has worked at the management level in capital equipment, military electronics, airframe and computer manufacturing. His consulting work includes productivity improvement and management education for Creative Output and General Electric clients. He holds an M.B.A. from Fairleigh Dickinson, and a B.S. from Adelphi University.

Dan Ciampa is President and Chief Executive Officer of Rath & Strong, Inc., of Lexington, Massachusetts. He frequently speaks and writes on merging total quality, just-in-time, and computer-integrated manufacturing; on involving people in technological change; on changing the organization climate; and, in particular, on the leader's role in continuous improvement efforts. He is the author of *Manufacturing's New Mandate* (New York: Wiley, 1988). He has been a guest lecturer at the Harvard Business School and at Boston College.

Mr. Ciampa received his bachelor of science degree in Finance and Social Psychology from Boston College and a Master of Education degree in Adult Education from Boston University.

He is a member of the Society of Manufacturing Engineers' Computer & Automated Systems Association, of Vanderbilt University's Operations Roundtable, and of the American Arbitration Association. In addition, he is an Associate of the British Institute of Management and a member of the Board of Advisers for *Corporate Controller*. Listed in *Who's Who in Industry and Finance*, Mr. Ciampa is certified by the Institute of Management Consultants and the International Association of Applied Social Scientists.

Romeyn Everdell is a management consultant, educator, and writer. In 1953 he joined Rath & Strong, working in industrial engineering, quality control, and production planning, scheduling and inventory control. He retired as Executive Vice President in 1985. His early industry experience included work as a quality engineer and production manager.

Mr. Everdell has been active in the American Management Association and the American Production and Inventory Control Society, as a frequent speaker and writer. He has also lectured for the American Society for Quality Control, the American Institute for Industrial Engineers, the Society of American Manufacturers, the University of Wisconsin, and Simmons College.

He wrote early technical papers in the areas of master scheduling and material requirements planning. APICS has honored his contributions to the field by establishing the Romeyn Everdell Award, given annually to the author of the best technical article in the *Production and Inventory Management Journal*.

He received his B.A. with honors in chemistry from Williams College.

Charles H. Fine teaches operations management and manufacturing strategy at MIT's Sloan School of Management. He holds doctoral and master's degrees from Stanford University, and a bachelor's degree from Duke University. He has authored a number of articles on manufacturing issues, including pieces on the economics of quality improvement and models for investment in flexible manufacturing systems. His industrial consulting, executive teaching, and research project experience includes work at Digital Equipment Corporation, Eastman Kodak, General Electric, IBM, and Motorola.

Stephen A. Hamilton is a senior consultant in HP's Strategic Manufacturing Consulting program where he focuses on the food process industry. Prior to joining the group, Mr. Hamilton was the project manager for the installation of major computer-integrated manufacturing (CIM) systems in Europe. He has held a variety of manufacturing management positions at HP in the United States, prior to which he worked for many years in public accounting in Europe and later the United States.

He earned his bachelor's degree in Politics, Philosophy, and Economics from the University of London and is qualified as a Chartered Accountant in England and as a Certified Public Accountant in the United States.

Ed Heard is President of Ed Heard & Associates, a manufacturing management counseling and educational services company based in

Nashville, Tennessee. He conducts SCM/JIT seminars and provides counseling services for a wide variety of major manufacturing companies and is the originator of short-cycle manufacturing.

Dr. Heard earned M.B.A. and D.B.A. degrees in Production Management and Industrial Engineering from Indiana University after working in materials management at Rockwell Manufacturing Company. Certified at the Fellow Level by APICS, he is an active member of APICS, IIE, and SME.

Julie A. Heard is Executive Vice President of Ed Heard & Associates, Inc. She has a background in manufacturing, accounting, and finance developed over the past 15 years. She has presented technical papers at International APICS Conferences and is a frequent speaker at APICS meetings, seminars, and regional conferences. She has served as APICS Vice President of Education Research, Region XI Vice President, and President of the Mid-Carolina Chapter in Columbia, SC. Ms. Heard studied accounting at the University of South Carolina and is certified by APICS.

William G. Holbrook is the Factory Manager for Stanadyne's Diesel Systems Division, which produces fuel system components.

He has been Vice President of Administration for the Stanadyne Automotive Products Group where he was responsible for financial reporting, data processing, purchasing, inventory, and production planning. The automotive products group has implemented a multiplant, repetitive production-planning system to support manufacturing processes and support functions for a seven-plant network. Stanadyne has used JIT techniques and systems for the past seven years.

Mr. Holbrook is a CPA with experience in public accounting as well as managerial accounting. He has served as a divisional and group controller. He is a frequent APICS speaker.

Frank S. Leonard is an independent management consultant specializing in the development and implementation of competitive business strategies. His consulting work includes acquisition studies, plant relocation and closings, and competitive manufacturing analysis. He has worked in production management and process engineering in several industries.

He has taught in M.B.A. and Executive Programs at Harvard Business School, Tulane University, and Simmons Graduate School of Management.

A frequent lecturer in the areas of strategy and operations management, his published articles have appeared in *The Harvard Business Review*, and Allan Kantrow's *Survival Strategies*.

Dr. Leonard's degrees include a B.S. in Chemical Engineering from the University of Missouri, an M.B.A. from Tulane University, and a Ph.D. in Business Administration from the Harvard Business School.

Jeanne Liedtka is Chair of the Department of Management of Simmons College where she teaches Business Policy and Strategic Management. Her industry experience includes work as a manager of Strategic Planning at Wang Laboratories and consulting in The Boston Consulting Group to clients on strategic planning as well as marketing, manufacturing, and distribution issues.

Her published articles appear in the *Journal of Business Ethics* and *The Proceedings of the Academy of Management*.

Dr. Liedtka holds a Ph.D. in Management Policy from Boston University, an M.B.A. from Harvard Business School, and a B.A. from Boston University.

Hal Mather is President of Hal Mather, Inc., Atlanta, Georgia, an international management consulting and education company.

Mr. Mather has authored *Bills of Materials, How to Really Manage Inventories*, and *Competitive Manufacturing*. His published articles have appeared in *Harvard Business Review* and *Chief Executive*. In 1987, he won the Romeyn Everdell Award for the best article published in the *Production and Inventory Management Journal*.

He has been certified at the Fellow level by the American Production and Inventory Control Society, is a Fellow of the Institution of Mechanical Engineers (U.K.), a Senior Member of the Computer and Automated Systems Association of the Society of Manufacturing Engineers, and is a member of both the Institute of Industrial Engineers, and the Association for Manufacturing Excellence. He is listed in *Who's Who in the South* and *Who's Who in Finance and Industry*.

Robert E. McInturff is President of McInturff & Associates, of Natick, Massachusetts, an executive recruiting firm specializing in materials management, manufacturing, and distribution. Prior to joining the company, Mr. McInturff had over seven years' experience in plant-and corporation-level materials management functions for Fortune 500 companies.

He writes a monthly column on career planning for *Electronic Buyer's News*, and co-authored a survey report titled "The Challenge and the Promise: Materials Management Today—An Evaluation."

He was a founding member of the Middlesex Chapter of APICS and served two years as its President. He is a frequent APICS speaker for local chapters and international conferences, and has lectured at Simmons College, Rhode Island College, Boston University, and the University of Massachusetts.

Mr. McInturff holds a B.S. from Northeastern University, and is recognized as a CPIM by APICS.

John W. Monroe is the Director of the Strategic Manufacturing Consulting Group at Hewlett-Packard. This group was created to give key customers access to the management techniques used to achieve significant operating improvements within Hewlett-Packard's own factories.

He has ten years of manufacturing experience at Hewlett-Packard. During his three years as Production Manager for the Data Terminals Division (now the Personal Office Computer Group), John was a member of the small team of managers who introduced Dr. W. Edwards Deming to HP management. He was also a General Manager at Avantek, Inc., which manufactures gallium-arsenide–based microwave components and subsystems for commercial and military users.

Dr. Monroe earned a bachelor's in Electrical Engineering, a masters in Electrical Engineering, and a doctorate, all from Cornell University.

Patricia E. Moody is a management consultant with over 15 years industry experience in materials management, production planning, and business planning, including positions with Digital Equipment Corporation, Data General, and 7 years with Rath and Strong.

She teaches operations management at Simmons College in Boston. She has spoken before international, regional, and local APICS conferences, as well as the Harvard Business School AMP Program and various industry seminars. Her published articles include pieces on systems analysis, distribution requirements planning, business planning, and service management. She is included in *Who's Who in the East*, is a member of the Professional Council, is certified by the Institute of

Management Consultants, and chaired the APICS Service Committee. She holds an M.B.A. from Simmons College Graduate School of Management, and a B.A. from the University of Massachusetts.

Roger W. Schmenner is an associate professor at the Indiana University School of Business. He has held faculty appointments at IMEDE (Lausanne, Switzerland), Duke, Harvard, and Yale Universities.

Dr. Schmenner's major interests lie in production and operations management. He is the author of the textbook, *Production/Operations Management: Concepts and Situations*, the editor of the casebook, *Cases in Production/Operations Management*, and the author of *Plant Tours in Production/Operations Management*. His research interests within the field include manufacturing strategy, productivity, and industry location. He has written over 50 published articles, book chapters, and cases that have appeared in the *Harvard Business Review*, *Sloan Management Review*, and *Journal of Operations Management*. Schmenner's book, *Making Business Location Decisions,* is a compendium of much of his stream of research on industry location. His recent research on factory productivity was sponsored by the U.S. Department of Commerce, Control Data Corporation, and IMEDE.

Dr. Schmenner has a diverse range of consulting and corporate teaching experience involving over 40 companies, several industry groups, and more than a dozen federal, state, and local government agencies or departments.

Dr. Schmenner holds an A.B. degree from Princeton and a Ph.D. from Yale in economics.

Mark Louis Smith is a an Industrial Engineering manager, and has held the positions of production manager, master scheduler, and strategic planning manager, all in the pharmaceutical industry. He has directed implementation of JIT and Kanban-based programs as well as MIS (management information systems) innovations, and employee empowerment projects. He holds a B.S. in Industrial Engineering from Bradley University and is completing an M.S. in Corporate and International Planning at the University of Pennsylvania.

Linda G. Sprague is Professor of Operations Management at the Whittemore School of Business and Economics, University of New Hampshire, and was Director of their Executive Programs, from 1981 to 1986. In 1984–85 she was Professor of Operations Management at IMEDE, the International Management Development Institute in Lausanne, Switzerland. In 1980 she was a Founding Professor at the National Center for Industrial Science and Technology Management Development at Dalian, China. She has also taught at Stanford University, The Tuck School, and Simmons Graduate School of Management.

Professor Sprague received her doctorate from the Harvard Business School; she also has an MBA from Boston University and an S.B. in Industrial Management from MIT. Her consulting and research interests include strategic management of operations, capacity management, and operations scheduling for manufacturing enterprises and for community general hospitals, production information systems, and productivity improvement programs. She has published articles on material requirements planning, international manufacturing, inventory management, and production practices in China.

Mrs. Sprague was Program Chair for the 1987 annual meeting of the Decision Sciences Institute of which she is a past president. She is a Vice President of the Operations Management Association and was Co-Chair of the 1987 Management Division of the Academy of Management. She is a certified practitioner in Inventory Management and a member of the Production Activity Control Committee of the Certification Council of APICS. She is a Fellow of the Decision Sciences Institute.

Professor Sprague is a member of the Board of Directors of Protek, Inc. Mrs. Sprague is the Management Advisor to SICOT, the Societe Internationale de Chirugie Orthopaedique et de Traumatologie.

INDEX